THE COMPLETE BOOK OF
OUTDOOR MASONRY

Other TAB books by the author:

No. 1080
$11.95

THE COMPLETE BOOK OF
OUTDOOR MASONRY

BY S. BLACKWELL DUNCAN

TAB BOOKS

BLUE RIDGE SUMMIT, PA. 17214

FIRST EDITION

FIRST PRINTING—MAY 1978

Copyright © 1978 by TAB BOOKS

Printed in the United States
of America

Library of Congress Cataloging in Publication Data

Duncan, S. Blackwell.
    The complete book of outdoor masonry.

    Includes index.
    1. Masonry—Amateurs' manuals. I. Title.
TH5313.D85   693   78-5046
ISBN 0-8306-9904-X
ISBN 0-8306-1080-4 pbk.

Cover photo courtesy of National Concrete Masonry Association

Contents

Introduction

Masonry is a term that covers a lot of ground. For all too many home mechanics and do-it-yourselfers, it connotes special skills, hard work and a field of endeavor that perhaps would be better left to professionals. But that's a shame, because the truth of the matter is that masonry work, in one or another of its guises, is well-suited to all kinds of do-it-yourself home projects.

Sure, there's some physical effort involved. There are skills (of a relatively low level) to be acquired and a modest amount of skull-work comes into play from time to time. But there's no reason to hang back; the same is true of carpentry, auto repair, cabinetmaking or just about anything else that you'd care to name. Masonry work is not difficult to learn about, nor hard to accomplish once you understand how to go about it.

There are a number of facets to masonry. It involves not only such popular materials as concrete, block and brick, but also less widely used elements like pavers, clay tile, ceramic tile, adobe and stone. This book covers all of these, and it delves into the mysteries of mortar, grout and stucco, too. Characteristics, specifications and uses of the various masonry materials are discussed, together with many of the design possibilities, construction methods and techniques employed in transforming the materials into useful, worthwhile and attractive structures.

Once you become acquainted with the fundamentals of masonry work, what you can do and what you can't do and how to tackle the actual construction on your own, the next step is practical application of your knowledge in projects around the house. With that in mind, the final chapters of this book are packed with more than three dozen general projects of a residential nature that you can readily put together. By introducing variations in design, materials and construction this series of projects can be augmented with several thousand more, because of the almost infinite flexibility inherent in masonry work. With thought and imagination backed by a basic knowledge of masonry, you can conjure up projects tailored to your needs and desires. The book's projects are meant to be exemplary only, and should be shifted about to suit your own situation.

Masonry is effective, efficient and extremely serviceable. Masonry structures are functional, rugged and of pleasing appearance, often of great beauty. They are relatively low-cost, require little maintenance and are easy to build. Masonry endures. What better reasons, then, to undertake some outdoor masonry projects around your own home?

S. Blackwell Duncan

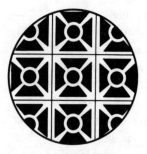

Acknowledgements

We wish to thank the many people who made this book possible. Special thanks go to the Brick Institute of America, the National Concrete Masonry Association, the Portland Cement Association, and Sakrete, Inc., for the use of their illustrations.

Tools and Equipment

As with any field of endeavor, there is a specific terminology used in, and which is peculiar to, masonry work. And as with any sort of building or construction process, certain tools and equipment are necessary to get the job done. Terminology will be treated throughout the book as the occasion arises. The tools and equipment needed, along with their use, will be discussed first, so that you will have a clear picture of what will be required, as we dig into the details of accomplishing all your own masonry work around your home.

SHOP TOOLS

To begin with, you will need an assortment of the standard carpentry tools most do-it-yourselfers are already thoroughly familiar with and have on hand. These tools are necessary for building forms, as well as for the many other incidental small jobs that are associated with masonry work. Such tools include a hammer or two, a carpenter's hand saw or an electric circular saw, a combination square or try square, a sliding T-bevel, and a flexible steel tape or carpenter's folding rule. A miscellany of wrenches, pliers, tinsnips, wire cutters, screwdrivers, wire brushes, and similar ordinary shop tools will come into play from time to time and be handy to have around. Both a chalkline and plumb bob (Fig. 1-1) will also see a lot of use.

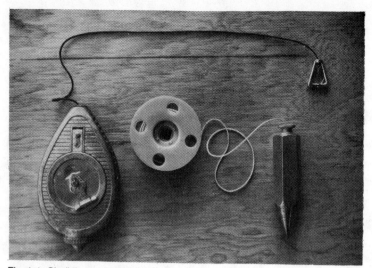

Fig. 1-1. Chalkline and chalkbox (left), and plumb bob with line and reel (right).

Bubble Levels

Levels are most important to this kind of work, and you will probably need at least three. One should be a small line level (Fig. 1-2) which can be attached to a taut line to measure grade elevations or to determine the level between two points. A short torpedo level (Fig. 1-3) can be used for leveling small sections, and a quality straight spirit level (Fig. 1-4) for larger sections. The latter level should be at least 2 feet long, and preferably 4 feet or even 6 feet, for checking the levels of slab concrete work, wall sections, footings, and the like. Long levels are also used to plumb walls and other tall structures. In addition, a small protractor level would be helpful on numerous occasions.

A Homemade Level

Often it is necessary to check the level between two points that are quite a distance apart. There are a number of ways to do this, but

Fig. 1-2. The line level attaches directly to a taut line or guideline. Courtesy of Stanley Tools.

Fig. 1-3. Small torpedo level.
Courtesy of Stanley Tools.

one of the easiest and most effective makes use of a level that you can cobble together yourself. Purchase a coil of clear or translucent plastic tubing. The length can be a standard coil of 25 or 50 feet, or you can use the particular length which suits your purposes best. The tubing size doesn't matter much, but 1/4-inch or 3/8-inch diameter will work fine. Make a pair of slightly tapered wood plugs for closing the ends of the tubing. Fill the tubing with water until the level rises to within a few inches of each end, and then plug the ends. Since water always seeks its own level, the top of the water column at one end of the tube will be exactly level with the top of the water column at the other end. To level two widely separated points, just lay the tubing out from point to point and adjust the grade stakes, level lines, or objects until they match the water levels in the tubing.

LAYOUT EQUIPMENT

Frequently one of the first acts in setting up for a concrete project is to determine exactly the topography of the site and to lay out the area. This means checking levels and grades, measuring out distances, setting proper right-angle corners, and establishing reference points so that the project will be placed in the correct position and everything will come out even.

Lines and Stakes

Much of this work is done with *taut lines* (which are merely stout strings run between stakes or posts), a line level to check the lie of the taut lines, and grade stakes. *Grade stakes* are generally

Fig. 1-4. Mason's 5-vial mahogany level (above), mason's all-purpose level (below). Courtesy of Stanley Tools.

Fig. 1-5. Split-image range finder is inexpensive and effective for shooting grades.

made from lengths of nominal one-by-two wood stakes, about two feet long, and sawed to a point at one end for easy driving into the ground. You can make up these stakes yourself, or buy them by the bundle at most lumberyards.

Leveling and Alignment Instruments

When the project covers a fair expanse of area, sometimes instruments must be used to determine correct levels and placements. The simplest of these instruments are the *carpenter's sight level*, the *pocket sight and surface level*, and the *Locke-type hand level*. Whatever the specific name, each is used in about the same way: you sight through them to determine rough grade and elevation. In many cases, they are quite sufficient for leveling, aligning and establishing site contours. Another instrument which works well is the Hoppe *split-image range finder* (Fig. 1-5). This is an inexpensive and useful gadget for determining levels and alignments, and can be used over distances from 3 to 300 feet. Accuracy is greatest in the lower distance ranges, but the overall results are excellent. This instrument is used by sighting through it to a special target, aligning the two images and then determining the elevations for alignment differential.

14

For extremely precise measurements, alignments, levels, and contour or elevation figures, you must resort to a *transit level* with tripod (Fig. 1-6) and *target rod*. Learning to use a transit for rudimentary work such as lining up fence posts or setting a level footing is not difficult, but the instruments are quite expensive. However, you may be able to rent one if necessary, and you can always hire someone with both the expertise and the instrument to do the layout work for you.

SITE PREPARATION EQUIPMENT

With most concrete work there is a certain amount of site preparation that must be taken care of before the actual work of forming and pouring can proceed. Often this work is concurrent with the site layout process. The tools used here are ordinary enough, and most likely you will have at least some of them around the house already.

Excavating Implements

A long-handled shovel or spade is essential, and a pickax or mattock may be necessary as well, for digging footing trenches, slab beds, leveling the ground, and the like. You'll need a steel garden

Fig. 1-6. Simple transit level for finding grades and elevations.

Fig. 1-7. Posthole digger and posthole bar.

rake for leveling earth, fill gravel, sand, and in some instances the freshly poured concrete as well. Setting posts and piers, especially small ones, can often be best done with a posthole spade, a posthole digger, and/or a posthole bar (Fig. 1-7). The bar is particularly worthwhile where the ground is tough and rocky. A crowbar or prybar also works well. A short-handled square-ended shovel is essential, not only for general shoveling work, but also for dry-mixing the concrete materials, hand-mixing small batches of concrete, feeding a power mixer, and for placing the wet concrete itself.

Tampers

One important aspect of readying a site for pouring concrete, especially in the case of slabs, is firming up and compacting the earth on which the concrete will lie. This is done with a *tamper* (Fig. 1-8). Once the surface is leveled, then the entire area is gone over with a tamper to pack the soil down hard and tight. Gravel added as a fill or subbase is compacted in the same fashion to obtain a relatively hard surface. Sometimes tamping will lower the grade level by virtue of the compaction, and then more fill must be added and the area tamped again. Occasionally, if the mix happens to be particularly stiff and dry, the freshly poured concrete itself must be tamped to achieve the proper density.

You can purchase or perhaps rent a tamper, but you can also build your own fairly easily. One possibility is to visit a local welding

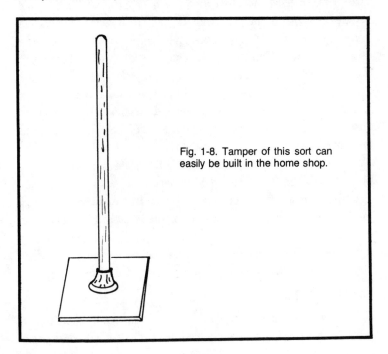

Fig. 1-8. Tamper of this sort can easily be built in the home shop.

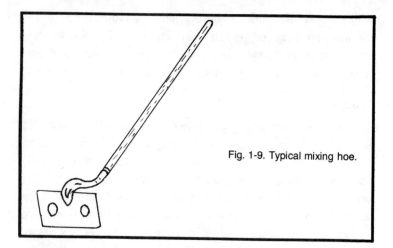

Fig. 1-9. Typical mixing hoe.

shop and have them torch out a 1-foot-square piece of heavy steel plate. Choose a fairly hefty piece that packs a lot of weight. Find the center of the plate by marking in the diagonals, and center a 2-inch pipe floor flange. Scribe the flange's bolt-hole pattern on the plate and drill corresponding holes. Countersink the holes on one side. Use 1/4-inch flat-head stove bolts of an appropriate length and secure the flange to the plate with the bolt heads facing down. Then screw a length of 2-inch pipe about 5 feet long into the flange to serve as a handle. This will make a fairly heavy tamper capable of flattening a considerable surface area in a short time.

If you are the rugged type and this tamper isn't heavy enough (remember, the more compaction you get, the better), then you can cast a small amount of concrete onto the top of the tamper. Drill and countersink a few more holes in the tamper face, set more bolts in place so that they will project above the plate surface. Make a shallow wood form to rest upon the upper surface of the tamper face. Pour the form full of concrete and let it cure. Be careful not to use too much, however, as concrete weighs on the order of 145 pounds per cubic foot. A tamper that you can't lift off the ground won't do you much good. An alternative method is to bolt extra pieces of heavy angle iron or chunks of lead to the top side of the tamper face.

MIXING EQUIPMENT

When it comes to actually mixing your concrete, a number of tools and pieces of equipment come into play. Exactly what you use will depend upon your specific situation, and you should choose whatever system seems to be the best for your type of project. The square-edged shovel has already been mentioned.

Hoes

A mixing hoe (Fig. 1-9) is sometimes used for mixing concrete, but more often for mortar, and thus is also known as a mortar hoe. Really, you can use a common garden hoe if you wish, but a standard masonry type does a better job and in less time. There are various sizes, but the one most commonly used has a head (or blade) about 10 inches wide and 8 inches high. Some types have a pair of large holes in the blade so that the ingredients can swoosh back and forth for better mixing.

Wheelbarrows

Mixing containers can take several forms. Probably the most popular is the wheelbarrow. For small batches of a couple of cubic feet or less, an ordinary small garden wheelbarrow works nicely. The somewhat larger utility barrow (Fig. 1-10) with a brim-full capacity of about 4 1/2 cubic feet, however, is probably the most practical all-around choice. Contractors' wheelbarrows of a 6-cubic-

Fig. 1-10. Typical utility wheelbarrow.

19

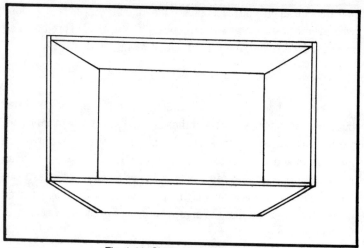

Fig. 1-11. Slant-sided mixing tub.

foot capacity are larger and sturdier, but considerably more expensive. Either of the latter two are also ideal for transporting concrete mix from the mixing site (or ready-mix truck) to the pouring site.

Containers for Mixing

Small batches of concrete can also be prepared in a *mixing tub*, the same as is commonly used for preparing batches of mortar. The tub usually consists of a small steel or wood tray, is watertight, and has slanted sides and a broad flat bottom for stability. Mixing can be done with a shovel or a hoe. Your can make your own tub from heavy tongue-and-groove planking, securely tied together, in whatever size you like. The wood variety is usually made with slanted ends and vertical sides (Fig. 1-11). Put the tub together solidly so that it can withstand the usual hard knocks, and make sure that it is watertight.

A *hoe box* (Fig. 1-12) works well for mixing small amounts of either concrete or mortar and is also easy to make. Again, make it rugged and watertight. Either mortar or cement can be easily mixed on a flat *mixing platform* such as the one shown in Fig. 1-13. A good size is about 12 feet square, allowing two men to work the mix simultaneously. Build a frame first out of two-by-fours, with plenty of crossbraces. Then plank the top of the frame with either nominal one-inch or two-inch boards, using matched stock such as tongue-and-groove. This will insure that the surface is watertight. Make the

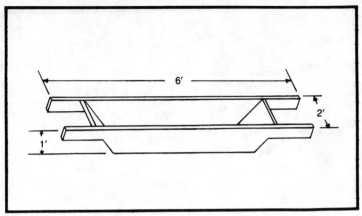

Fig. 1-12. Hoe box for mixing concrete or mortar.

whole affair sturdy, since the platform will be subjected to a considerable amount of weight.

Power Mixers

The easiest way of mixing concrete or mortar, of course, is with a *power mixer* (Fig. 1-14). A wide range of sizes and styles is available from a number of different manufacturers, and while some models are extremely expensive, others are well within the financial reach of the do-it-yourselfer who contemplates working on a number

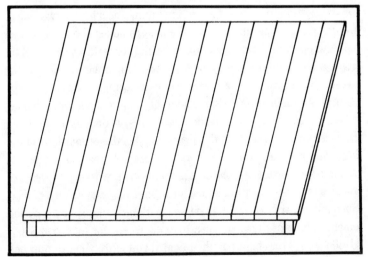

Fig. 1-13. Wooden mixing platform.

Fig. 1-14. Large side-dumping power mixer, a type widely used for mixing mortar and stucco. End-dumping drum type is popular for mixing concrete.

of masonry projects over the years. Like any piece of large equipment, a mixer is an investment that has long life and pays good returns.

Mixers are available in stationary, portable or mobile styles. The stationary type has to be muscled into position and remains there on its solid base. The portable type is equipped with rubber-tired wheels of the so-called utility or low-speed type, so that the machine can be readily trundled about the site or transported in the bed of a trailer or pickup truck. The mobile type is equipped with large wheels capable of highway travel, and is designed to be towed behind a vehicle from job to job.

Capacities of these mixers are often specified in two ways; maximum, and operating. Since the mixers never should be loaded full, the operating capacity is the specification with which you'll be concerned. Capacities may be expressed in either cubic feet or sacks. (Standard sacks of portland cement contain 1 cubic foot—or 94 pounds—of material.) On the average, 1 sack of portland cement, when combined with water and aggregates, produces about 4 1/2 cubic feet of concrete, but this depends upon the proportions of the ingredients. One of the most common sizes of small power mixers has a 1-sack drum, absolute capacity, with a working capacity of about half a sack, to yield batches of about 2 1/2 cubic feet of mix. Such mixers are powered by either an electric motor or a gasoline engine. Some of these rigs can be bought new for less than $200. Used rigs sell for about half the price. If you wish, you can rent one from a large tool and equipment rental center.

There are some odds and ends that you will need during the mixing process, too. If you are screening your own sand, then obviously you will need a *sand screen* (Fig. 1-15). This is simply constructed by nailing together a rectangular frame of whatever size is convenient from two-by-fours or two-by-sixes and fitting it with a couple of angled legs at one end so that the screen stands at an approximate 45-degree angle. Cover the bottom of the frame with screening of an appropriate mesh size, securely nailed or stapled in place.

Proportioning Equipment

The *bottomless box* (Fig. 1-16) is also easy to make, so much so that you might like to have two or three different sizes around for

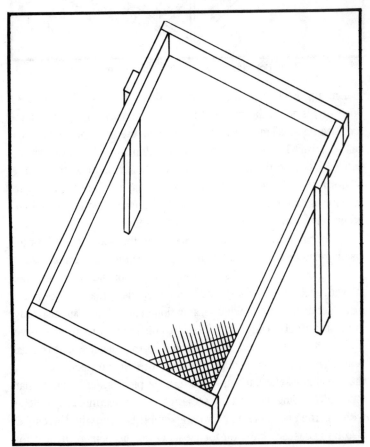

Fig. 1-15. This type of sand screen can be easily built in the home shop.

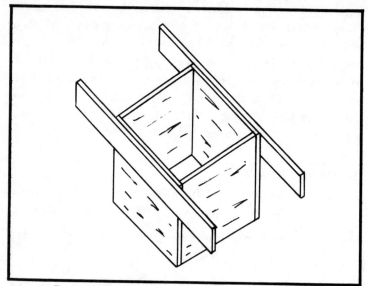

Fig. 1-16. Bottomless box for portioning dry-mix materials.

making up batches of varying proportions. The 1-cubic-foot size, however, is about the handiest for concrete work, and a 1/2-cubic-foot size for small mortaring jobs. By using the same principles, you can construct 1 1/2-, 2-, or even 3-cubic-foot sizes for mixing larger batches. To use these boxes, you simply set them on the mixing platform or in the wheelbarrow or whatever, dump in the requisite amount of material, and then lift the box away, giving yourself an accurately proportioned quantity with no fuss.

Mixing concrete requires water, and the water must be precisely measured. For small batches, you can use a 1-gallon plastic jug, with graduation marks registered on the outside in waterproof ink with a felt-tip pen. Plastic pails can be similarly marked; some are manufactured with molded graduation lines. You can also buy graduated steel buckets, but these are considerably more expensive than plastic ones. By having a word with the management of local restaurants you may be able to obtain, free for the asking, a few of the 5-gallon plastic buckets used to pack pickles, lard, and similar foodstuffs. Filling the water vessels at the job site is most easily done with a garden hose equipped with a shutoff-type nozzle. Hoses will also be necessary later on for cleaning up the equipment and for moist-curing the concrete after it has been poured and finished.

24

EMPLACEMENT TOOLS

Some of the equipment you'll need during the pouring phase of the project has already been mentioned. The wheelbarrow can be used to transport the mix and square-tipped shovels to move the mix into place. A steel garden rake turned upside down works well for bringing the poured mix to a rough level.

Puddling Tools

A square-tipped shovel can also be used for the procedure known as *spading* or *puddling*. This consists of sticking the shovel blade straight down into the mix (whether the pour happens to be a slab or a thick section, such as a footing) and wiggling the blade

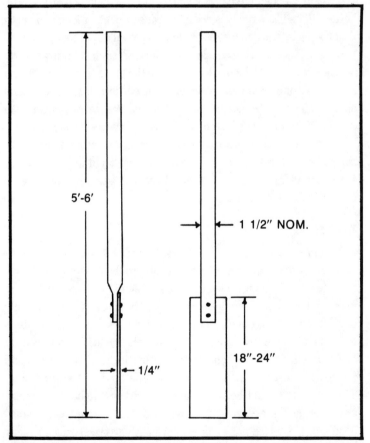

Fig. 1-17. Puddler which can be home-built from scrap pipe and steel strap.

25

around to help the mix flow out, settle the ingredients into a fairly even consistency, release entrapped air bubbles, and improve the overall density of the mix.

Spading is a particularly important maneuver along the edges of the forms, where large chunks of aggregate tend to collect against the form walls and present a rough surface when the forms are stripped. The shovel blade is inserted directly adjacent to the form side and wiggled about, so that the larger pieces of aggregate are impelled away and are replaced by smaller particles and "soup." A shovel blade is only effective, however, in shallow pours. If the concrete mix is deep, as in a fully-poured wall form, then you will need a *puddler* (Fig. 1-17). If you don't have too much area to cover, you can use a posthole bar in a pinch. The posthole digger is a little better but its blade really is not quite wide enough or long enough to do a good job. You can make a puddler yourself out of easily obtainable materials. The blade consists of a piece of tough steel strap about 4 inches wide and 18 inches long. Attach the blade to a length of pipe about 4 1/2 or 5 feet long, flattened at one end in a vise or with a hand sledge. The pipe size doesn't make much difference, but a trade size of 1 1/4-inch or 1 1/2-inch diameter affords a good hand grip.

To use a puddler, slide the blade deep into the concrete alongside the form wall and work the tool back and forth, using the form as a fulcrum, until all the visible coarse aggregate disappears into the mix, leaving a smooth surface.

The Jitterbug

There are times when a *jitterbug* (Fig. 1-18) is necessary to bring the surface of a poured concrete slab to the proper consistency for effective finishing. The jitterbug is actually a special kind of tamper; it is used by repeatedly bumping the tool against the freshly poured concrete surface. The purpose is to bring enough moisture and cement to the top to permit a creditable finish. This tool is used only in cases where the concrete mix is particularly stiff and dry and no moisture rises to the surface during the pouring and spreading. The jitterbug should be used carefully and sparingly, and, unless absolutely necessary, should not be used at all. There are also small hand tampers which are employed primarily to cover smaller areas than the jitterbug covers. These tampers are used under the same

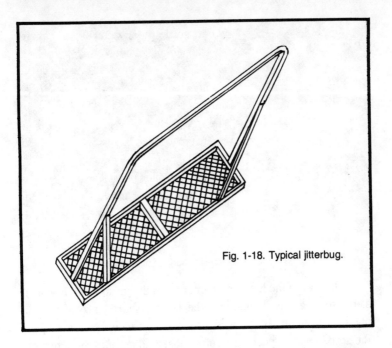

Fig. 1-18. Typical jitterbug.

circumstances and for the same purpose—to compact and make ready for finishing an exceptionally stiff and nearly unworkable mix.

FINISHING TOOLS

After the concrete has been poured and leveled, the first finishing step is called *screeding* or *striking*.

The Screed

The tool used for this purpose is called either a *screed* or a *strike-board* (Fig. 1-19). The screed is a straightedge which is set across the top of the form and then pushed forward with a back-and-forth sawing motion, scraping the surface of the mix level and smooth. The excess mix is pushed ahead of the screed and worked into any low spots along the way. In most instances, the screed is perfectly straight, in order to make the concrete surface as level as possible. Best results are obtained by using a tool which is manufactured as a straightedge. More often than not, though, the screed consists of the nearest handy piece of lumber that has one edge free from defects and appears relatively straight to the eye.

Fig. 1-19. Screed for leveling surface of fresh concrete.

The size of the lumber depends upon the job at hand. For small slabs, footings, wall tops and the like, a short piece of nominal one-by-four or one-by-six will do the job. For wider spans and where the screed will be pushing a substantial amount of excess concrete ahead of itself, then something stiffer must be used, such as a length of two-by-four or two-by-six or even larger, in order to avoid bowing and sagging. Unless the job is quite small, screeding is a two-man job, since the screed is hard to handle and does not push along readily.

In certain instances, it is desirable that the screed's bottom edge not be straight, but curve slightly downward or upward, or be just slightly V-shaped. The screed can be made in this shape by trimming to the desired curve with a power plane, or by deliberately choosing a bowed piece of dimension-stock lumber to serve the same purpose. The application is most generally on driveways where a slight crown is built into the finished surface to aid in water drainage. In some cases, the crown is inverted, or slanted into an extremely shallow vee, so that moisture will drain off down the center of the drive. Crowning is also sometimes used on walkways, for the same purpose.

The Darby

As soon as the screeding is finished, the next step is to treat the fresh surface with a *darby* (Fig. 1-20). The darby is a flat piece of wood, magnesium or aluminum, anywhere from about 2 1/2 to 7 feet long, and usually about 4 inches wide. A long, angled handle is attached to the top surface, so that the tool can be floated across the surface of the concrete, using little or no downward pressure. This smoothes out the serrations and ripples left by the screed; it also fills in any small low spots and knocks off high ones. At the same time, bits of coarse aggregate which might be close to or protruding from the surface are sufficiently embedded in the mix to allow a proper finishing job. In fact, the darby may also be used on small sections as a screed. Thus two processes are taken care of with one tool, and the operator is enabled to make as few passes across the fresh surface as possible.

The Float

The first stage of the actual finishing process, which generally takes place an hour or so after darbying, depending upon the specific

Fig. 1-20. Darby for finish leveling of fresh concrete surface.

Fig. 1-21. Typical wood hand float.

curing conditions, is accomplished with a *float* (Fig. 1-21). There are several types of floats, but they all are used for the same purpose and in much the same manner. The choice of which one to use depends upon the extent of the concrete surface to be finished, and upon the desired finish texture. Floats, which come in numerous sizes, are made from either wood or metal. A wood float produces a slightly rough finish, while a metal float will produce a finish anywhere from smooth to slick.

The most elementary float consists of a short chunk of scrap two-by-four with a flat and smooth face. The edges and corners are rounded off so the tool will not dig into the concrete surface. This simple float can be anywhere from 6 inches to a foot long and, if carefully worked, will give good results. Perhaps the most common type of wooden float consists of a board, often hardwood, about 4 inches wide and a foot long, with a full handle attached to the top surface for ease of working.

Metal hand floats are made in much the same manner, and may be fashioned from aluminum, steel, or magnesium. All of these floats are used from a kneeling position, and will cover only a small area at a time.

For larger areas, where the surface may be difficult to reach or not yet hard enough to get out onto with kneeling boards, or where working large areas in a short time is desirable, then much larger long-handled floats are used (Fig. 1-22). A simple wood float consists of a flat, smooth board, from 3 to 6 or 7 feet long and from 4 to 6 inches wide. The board is attached at an angle to a long pole handle

Fig. 1-22. Long-handled float.

and securely held in place with braces. This type of float can be made up right on the job site, and does a remarkably effective job if handled properly. There are also any number of commercially made metal floats, either steel, aluminum or magnesium, which are similarly fitted out with long handles. It is also possible to use a darby for floating relatively small surfaces.

The floating process is done by drifting the float delicately across the concrete surface. No pressure is applied to the float, save for the small amount sometimes needed to smooth out ripples or high spots. Care must be taken to keep the float surface almost level, but not quite, so that wavy lines or chattering spots will not appear behind it as it is moved; but on the other hand, the float's leading edge must not be allowed to dig into the concrete surface. The float should be moved smoothly forward and back or side to side, overlapping slightly on each stroke but eliminating at the same time any lap marks or slight ridges from the previous pass. The purpose of floating is to bring just enough moisture and cement particles to the surface to form a dense and compact mass which will have great strength when the curing is completed. Floating also accomplishes the final leveling of the surface.

The Trowel

If an even denser and smoother surface than can be obtained by metal floating is desired, then the tool to use is a *finish trowel* (Fig. 1-23). The finish trowel looks much like a metal hand float, except for a somewhat different handle design in most cases. In fact a metal hand float is often used for finish troweling. A proper finish trowel, however, is usually manufactured from spring steel which can flex to a certain degree. It is made in various sizes, with a large one being about 5 inches wide and 12 inches long, and the sizes graduating downward from there.

Finish troweling is done after the floating phase has been completed, and the concrete has been allowed to set up and harden further. Unlike floating, finish troweling is done by applying varying degrees of pressure to the trowel blade to make a smooth, slick, compact and dense surface. One pass over a floated surface will result in a smooth finish. Subsequent passes will render the surface even smoother, until finally no further progress can be made and the

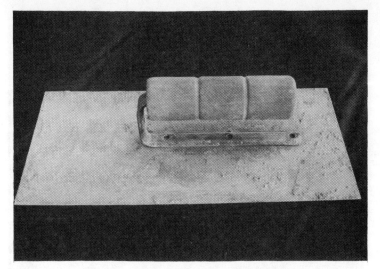

Fig. 1-23. Steel finish trowel. Duncan Photos.

surface is so smooth and blemish-free that it has a glazed appearance.

Often it is necessary for the worker to get out onto the fresh concrete surface to do his finishing work. For this, one or several *kneeling boards* are required. Chunks of plywood cut to about a 2-foot or 3-foot size work well for the operation. In any case, the boards should be large enough so that the worker can kneel comfortably, have room on the board to brace the heel of one hand while he works with the other, and be long enough so that his toes do not slip off the end and dig into the concrete. The corners and lower edges of the board should be rounded to reduce the possibility of gouging the concrete surface. Never attempt to walk on fresh concrete; move from kneeling board to kneeling board as necessary. The point at which the concrete is sufficiently set up to receive the weight of the worker is an open question. There is no particular waiting period and no rule that can be applied; all you can do is test the concrete and judge whether the surface is firm enough.

Groovers and Edgers

Another tool often used in finishing work is a *groover* (Fig. 1-24). *Grooving* is primarily done to sidewalks, but may also be done to driveways. The grooving tool consists of a handle attached to a

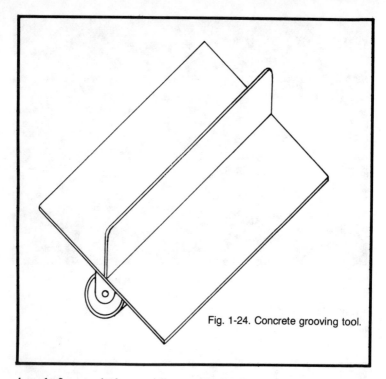

Fig. 1-24. Concrete grooving tool.

broad, flat steel plate, with a stubby blade protruding downward from the center of the plate. The blade creates a groove about an inch deep in the fresh concrete, while the flat portion forms a characteristic smooth band along each side of the groove. Though grooving can be employed to achieve a decorative effect, the main reason for its use is to provide periodic weak points in concrete slabs. Then, if the slab is beset by irresistible stresses of frost, tree roots, subsoil settling or the like, any cracks that develop will occur unobtrusively along the grooves and not randomly along the slab length.

An *edger* (Fig. 1-25) is usually used in conjunction with a groover, although *edging* may be done alone. The edging tool is little more than half of a grooving tool, except that the blade is placed to one side and is gently curved along its short axis. When run along between the form rails and the fresh concrete of a drive or walkway slab, the edger generates a slightly rounded edge with an attractive ribbon-like margin alongside. Although edging is sometimes done for decorative purposes, the main reason for its use is to mold an edge to the slab which is less susceptible to crumbling, nicking, or chipping away.

Brooms and Brushes

One other commonly used finishing process employs *brooms* to gain the desired effect. There is no need to purchase special brooms (unless you wish to); the job can be done with practically any kind of broom or brush, depending upon the size of the area and the desired effect. An ordinary household or kitchen broom works fine, provided that the working edge is clean and square, and not rounded off or scraggly. For small jobs where all parts of the pour are easily reached, you could use a fireplace broom, a dustpan brush, a whisk broom, a scrub brush, a mechanic's wire brush, or any one of a hundred others. You can even draw designs with an old toothbrush, if you feel up to a tedious job.

Larger surfaces are best done with a push broom, such as a shop broom or a street-sweeping broom. Assorted combinations of bristle lengths and stiffness will produce different patterns on the concrete surface. The trick is to keep the lines even and straight, with a minimum of discordant striations and overlapping. The brush lines normally are run at right angles to the traffic flow, but you can

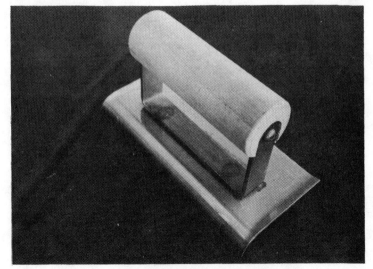

Fig. 1-25. Concrete edging tool.

Fig. 1-26. Masonry trowel, also called general-purpose, mason's or brick trowel.

also introduce patterns such as squiggles, chevrons, and her-ringbones.

UNIT MASONRY TOOLS

Working with masonry units, such as concrete block, brick, pavers, and a number of other similar modular building blocks, requires many of the tools already mentioned, plus a few more. More often than not the units are bonded together with mortar, which must first be mixed. For this you can use a *mortar tub*, which really is no different than a concrete mixing tub, or you can use a hoe box, a mixing platform, or a wheelbarrow. The mixing is done with a shovel or a hoe. Hoses, buckets, and the rest of the odds and ends are the same as for concrete.

The Mortar Board

When the mortar mix is ready, it is usually shoveled onto a *mortar board* in small quantities. Commercial types of mortar boards, which are made from wood or steel, and sometimes are even electrically heated for cold-weather work, can be purchased at building supply outlets. However, probably the most commonly used material for mortar boards is a chunk of fairly heavy plywood, either scrap wood or otherwise, with one unflawed surface. This seems the most reasonable type for the do-it-yourselfer, and a size of about 2 feet square should be ample. The board is positioned right next to the working area, and a bucketful of mortar dumped on top. The mortar is then scooped up as needed, with the pile on the mortar board being constantly replenished from the mixing tub.

Mortaring Trowels

Application of the mortar to the blocks or bricks or whatever is done with a *masonry trowel*, sometimes called a *brick trowel* (Fig. 1-26). There are a number of patterns which vary only in minor design details, but most embody a tempered steel blade, which is a bit springy, and a well-secured hardwood handle. The trowels vary in width from about 4 1/2 to 6 inches, and are anywhere from 8 to 14 inches in length. Generally speaking, the larger the masonry unit being mortared, the larger the trowel: for concrete blocks a large size, for bricks a smaller size. However, much depends upon personal preference; most masons have their own likes and dislikes, and indeed usually have a favorite trowel from which they will not part. The masonry trowel is used for virtually all kinds of ordinary masonry work.

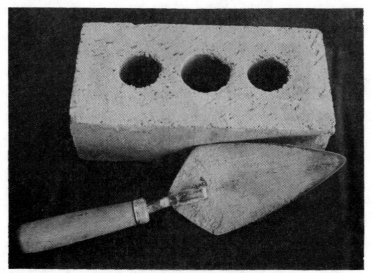

Fig. 1-27. Small buttering trowel. Note rounded tip.

Fig. 1-28. Large pointing trowel. Note pointed tip.

When working with smaller units such as bricks, however, many masons prefer to use a *buttering trowel* (Fig. 1-27). There are two principal differences between the buttering trowel and the masonry trowel. Though the physical proportions of the two types are about the same, the overall size of the buttering trowel is generally smaller. More importantly, the tip of a buttering trowel is well rounded off, not pointed like a masonry trowel tip. The principal use is in laying the mortar on bricks, a process widely known as buttering.

Pointing trowels (Fig. 1-28) are more specialized than the other two types of trowels, and they are also considerably smaller and lighter. They may run as small as 2 1/2 inches wide and 4 inches long, or as large as perhaps 5 1/2 inches wide and 7 inches long. There are a lot of different shapes for various specific purposes, but the most common design is almost identical with that of the masonry trowel. Although it is sharply pointed, that's not the reason for this trowels's name, however. The trowel is used for finishing mortar joints after the laying process is complete, a task known as *pointing up*. The same trowel is used for repairing old mortar joints which have begun to crack apart and crumble out, in a process called *tuck pointing*.

Hammers and Chisels

Frequently masonry units must be cut or reshaped somewhat in order to fit around other structural elements, or to make the courses come out even, or for a number of other reasons. Here you have several choices of tools, but you have to be guided somewhat by the nature of the masonry unit itself. The *mason's hammer* (Fig. 1-29), also called a *brick hammer*, is a utilitarian tool. There are several styles, sizes and weights, and there is a knack to using one which requires practice or experience. One end of the hammer head can be used for setting stakes, driving nails, tapping chisels, or for making rough breaks in block or brick. The opposite end, the *chisel peen*, is used for trimming rough cuts or breaks in the masonry units. For best results, the hammer head should be kept dressed flat, and the peen kept sharp.

The *scutch* (Fig. 1-30) is a tool vaguely similar in appearance to a brick hammer, except there is no head. Instead, a cutting peen extends to either side of the handle, much like a flat-bladed pickax in miniature. This tool is particularly used for trimming and cutting brick, though it could be used for various types of concrete block work as well. But unless you anticipate doing a lot of hand cutting, or just like to collect good tools as many do-it-yourselfers do, you probably will not need one.

Using a *mason's chisel* (Fig. 1-31), also called a *blocking chisel*, is probably the best way to cut bricks or blocks by hand. This is a short, wide-bladed chisel with a stubby and rugged metal handle (the striking shaft), all made in one piece. There are a lot of weights and sizes and slightly different designs, and which one you choose matters little, since they are all essentially the same. The type with a broad bevel on one side of the blade is perhaps the most popular.

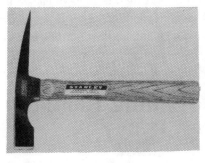

Fig. 1-29. Mason's hammer, also called a brick or bricklayer's hammer. Courtesy of Stanley Tools.

Fig. 1-30. Scutch, used for cutting brick.

To use the chisel, you position the blade with the bevel facing you, and the section of brick or block being cut away facing in the opposite direction. A few stout raps on the chisel will do the job. Or, you can first score the material with a few light taps and then use the mason's hammer to break the piece along the scored line. But don't use the mason's hammer to do any hefty striking. A heavy wood *maul* or *beetle* works well and is safest, since there is no danger of metal shards breaking loose from the striking point of the chisel handle. If this utensil fails, use a hand sledge with a 2-pound to 4-pound head. Incidentally, the sledge is useful for driving grade stakes and form-support stakes, and also for rock work.

Cutting masonry units by hand is not an easy job, does take time, and requires a certain amount of skill. During the process of developing that skill, you can consume a lot of bricks or blocks by smashing them to flinders instead of cutting them properly. If you have much cutting to do, an alternative course is to rent or borrow a *power masonry saw* (Fig. 1-32). If you can't do that, you might arrange to have a local contractor make the cuts for you during some slack time, provided that you transport the materials to him. The power masonry saw, though somewhat dangerous to use, does indeed make fast, neat and clean cuts that are a joy to work with as

Fig. 1-31. Mason's or block chisel. Courtesy of Stanley Tools.

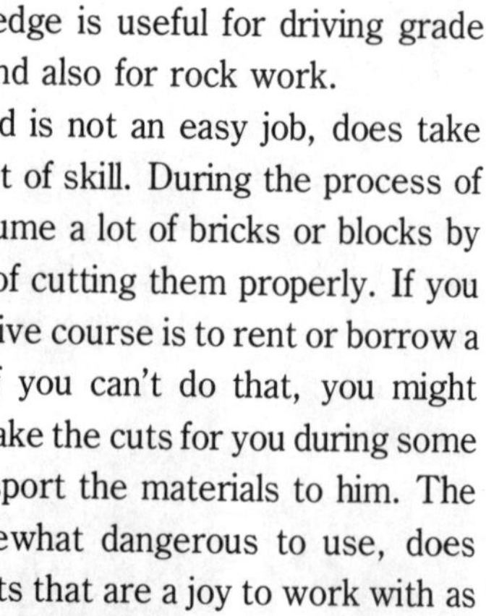

the structure is assembled. But you would have to do a considerable amount of cutting to justify the purchase of one for home use.

The Mason's Line

A *mason's line*, or *taut line*, is an essential bit of equipment in nearly all masonry unit construction. The line is pulled tight, so that there is no sag, between the corners (or from end to end) of a wall structure. The line may be anchored by tying the end to a block or brick or some fixed object at each end of the course, or may be attached to *mason's adjustable line holders* which are locked onto the structure. Actually, any sort of anchoring points that will hold the lines perfectly tight, will allow no sag, can be readily and easily moved, and is not likely to be jostled out of position, will do the job. The line itself serves both as a level line and as a reference point. As each course of masonry units is laid, the individual units are aligned with each other, with the course below, and with the taut line above. The line is then moved to the next higher course, and so forth. Used along with a spirit level, the taut line insures that the structure will remain plumb and true through to completion of the project.

Jointing Tools

When a section of masonry unit construction is completed, or sometimes during the course of its construction, the mortar joints

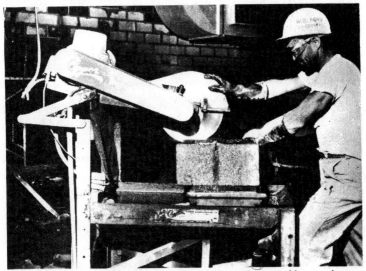

Fig. 1-32. Power masonry saw. Courtesy of National Concrete Masonry Assoc.

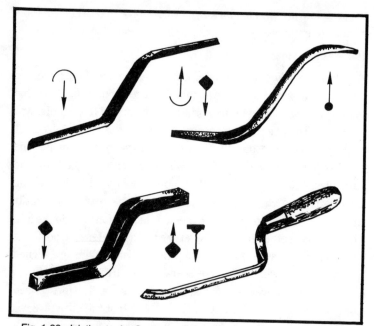

Fig. 1-33. Jointing tools. Courtesy of National Concrete Masonry Assoc.

may be finished simply by striking them. This involves scraping off any excess mortar with the edge of the trowel. The result is a flat joint which lies flush with the face of the masonry unit. Sometimes, though, another type of joint may be desired. This could be just for effect or eye appeal, or because further compaction of the joint is desired, or because measures to prevent moisture penetration into the joint are deemed necessary.

In some cases, a pointing trowel can be used to change the joint form. *Jointing tools* or *jointers* (Fig. 1-33) are also often used, and these can be acquired in a number of sizes and shapes. In fact, many masons make up their own jointers from pieces of iron rod, with the flats or tips filed or ground into whatever joint shapes are desired. Some of the more common shapes used are convex, concave, doubled or raked, and V-joint. Actually, you can use practically anything as a jointer if it is solid enough, will withstand the constant abrasion well, and is shaped to suit your needs. The main object is to compress the joint as much as is reasonably possible, leaving a smooth, dense finish in whatever particular pattern you have chosen. This procedure of jointing is sometimes called *tooling the joints*.

STONEWORKING TOOLS

When stone, rather than premanufactured masonry modules, is used for unit construction, some further equipment may be needed. This is not always the case, because flagstones and some types of veneer stones may either be fitted in mortar bedding in random sizes or pretrimmed into regular shapes which are easily assembled. Rough or native stone, however, is another matter. In the rubblestone (fieldstone) type of construction, the rocks must often be sized and trimmed to fit.

Collection and Movement of the Stones

The first problem in working with native stone is in collecting the stone. In some areas you can purchase suitable rock by the ton from a rock supplier, in which case you can have the materials delivered to you and deposited right at the job site. Often, though, the stonemason prefers to collect his own materials. This is done in whatever manner seems handiest, with a pickup truck or a tractor and stone sledge. Moving the rock about at the job site can be done with a wheelbarrow, but a large, stout garden cart with cycle-type wheels (Fig. 1-34) works even better. Heavy stones can be loaded into any conveyance by rolling them up a stout plank studded with 1-by 2-inch wood strips nailed crosswise at intervals, ladder-fashion. The strips hold the rock in place while you catch your breath for the next heave. Large stones are also run into position atop a wall or crib in the same manner.

Fig. 1-34. High-wheeled garden cart. Courtesy of Garden Way Research.

Fig. 1-35. Besides the traditional square-headed stonemason's hammer, either the New England pattern blacksmith's hammer (above) or the hand drilling hammer (below) works nicely. Courtesy of Stanley Tools.

Personal Safety

A word about safety is important here. When you set to work shifting native rock around, either in the collecting or in the building phase, strange things sometimes happen. Native stone is extremely heavy, awkward to handle, often sharp-cornered, highly abrasive, and unstable. This means that you have to work carefully if you value your fingers and thumbs and shins and toes, because rocks sometimes go their own way.

A pair of steel-toed work boots is really a smart idea. At the very least, wear heavy box-toed work or hiking boots and never sneakers or light shoes. A mashed toe is terribly painful. A pair of heavy leather gauntlet gloves is essential if you want to avoid numerous cuts and scrapes and blisters. Whenever you are setting stone, especially the larger sizes in dry-laid construction, wear those gloves. That simple precaution will save a lot of flattened fingertips and thumbs.

When you are breaking rocks, even little ones, wear a face shield, or at least safety goggles. Either device should be shatterproof. Those shards of rock flying from broken stones strike hard and fast, and can be extremely dangerous, especially to the eyes. The fragments will penetrate your flesh like Indian arrowheads if they hit just right, and there really is no satisfactory way to avoid them.

Tools for Altering and Installing Stones

Breaking up large rocks or boulders requires the use of that all-time favorite, the long-handled *sledgehammer*. The 8-pound size is the easiest to swing, but will just bounce off a really tough rock. For better results, move up to the 10-pound head, or even to the 12-pounder. Beyond that, you're likely to do more damage to yourself than the rock.

Small stones are most easily broken with a *stonemason's hammer* (Fig. 1-35). These run from 2 to 4 pounds or so in head weight and have handles about 10 to 16 inches long. One side of the head is ground flat with a beveled edge, while the opposite side is wedge shaped. For light cracking and picking work, as for shaping wedge stones or for light trimming, a *geologist's* or *rock-hound's pick* (Fig. 1-36) works well. These implements are available in several sizes and styles, with weights ranging from about 1 to 1 1/2 pounds or so. The *bush hammer* (Fig. 1-37) is a tool that is peculiar to stonemasonry. It looks rather like the meat-tenderizing mallet the butcher uses, with a big, square head that is toothed on both faces. The standard brick hammer is also a useful tool in stonework.

Cutting and trimming stones can also be done with chisels, and a great variety of appropriate models is available, especially if you want to do stone sculpturing. However, for roughing out masonry units, the blocking chisel mentioned earlier will do a creditable job. In addition, there are various *plain chisels, toothed chisels, pitching*

Fig. 1-36. Geologist's hammers, or rock-picks.

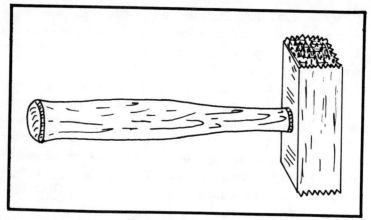

Fig. 1-37. Bush hammer.

tools, and *points* which are most useful. There are many designs, sizes, shapes, and weights, but they all are about equal in effectiveness and your choice should be governed by what is available to you or what style you happen to like. Almost any broad-bladed masonry chisel with integral steel handle and striking surface will do a fine job once you learn the techniques, and provided you keep the blade sharp. Points (the implements) are used for picking work, while pitching tools are just a toothed variation of the chisel. For drilling, there is an assortment of *rock drills, star drills, miner's drills* and so forth.

Splitting or cleaving large stones (as opposed to merely breaking rocks or light hand quarrying) requires a few more odds and ends. These include heavy *wedges* and *wedge-and-shim sets* for forcing the stone layers apart, and a *crowbar* or two to help the process out. An array of sledges and hammers is also needed. Be aware, though, that this type of work is time-consuming, difficult, and exceedingly laborious. Hand quarrying can also be extremely dangerous, and should never be done alone. If you feel energetic enough to investigate this source of good stone, seek out local advice and assistance, and don't jump in whole-hog before you know what you're about.

Setting and adjusting stones in dry-laid work, and sometimes in mortared construction as well, can be a finicky job because the stones don't always want to rest exactly as you'd like. Inevitably there is an amount of shifting and tugging and adjusting and wiggling to be done. Levering medium-weight stones into place can be accomplished with a length of two-by-four or four-by-four, or a long pole cut from a green sapling. A broad-bladed post-hole digger will do a better job. For large and heavy stones, you will need at least one 5-foot or 6-foot wedge-tipped crowbar, possibly two. Making slight adjustments of position once the stone is set in its approximate location is much better done by inserting a steel bar in the crack than your fingers. A standard 24-inch or 30-inch *wrecking bar* works nicely for this, and so does the smaller *rip bar* or a stubby but thin-bladed *pry bar*. If you have some old iron pipes on hand, or steel tubing, you can make a useful lever by flattening one end with a sledgehammer.

TILE-WORKING TOOLS

Various kinds of tile are sometimes used for outdoor projects, and working with them may necessitate the addition of a few more tools to your kit. Ceramic tile, for instance, can be cut by scoring with a *glass cutter* or a special *diamond-tipped cutter*, and then breaking. There are various kinds of *tile cutters* that produce a neat, trim edge, and these you can either purchase or rent. *Nibblers*, of which there are many styles, sizes, and shapes, are also quite useful for cutting tile to fit around openings, pipes, or what have you. Special saw blades, like the Remington "Grit-Edge" blades, can be obtained in special rod saw form, hacksaw form, and sabersaw form. These blades have a cutting edge consisting of tungsten carbide particles and will make straight or contour cuts with equal ease—it's almost as though you were working with wood. These same blades can be used to cut either natural or synthetic marble.

The Concrete Facts

The outdoor masonry projects that you might undertake around your home can be roughly broken down into three categories.

DRY UNIT MASONRY

The first category is the so-called "dry" masonry, or "dry-laid" masonry, which requires no bonding agent between the elements of the structure. This would be the case where stones are simply piled symmetrically to form a wall, or where a row of bricks is partially buried in the ground to form a garden or walkway edging. None of the unit pieces of the project are tied to each other, but merely depend upon the support of soil, balance, dead weight, and the inherent friction of the surfaces which rest together to form the completed structure. Obviously, this type of construction is neither sturdy nor solid enough for most projects and has little rigidity, strength, or durability. The method is unsuitable for many projects. On the other hand, dry-laid masonry is easy to do and requires little in the way of materials and equipment. Still, there are certain projects which lend themselves very nicely to this technique, and are in fact longer-lived than when built by other means.

UNIT-AND-CEMENT MASONRY

In the second category, the individual building blocks of the projects *are* tied together, piece by piece, with a cement. There are numerous kinds of masonry units which can be assembled in this fashion, the most common being brick, concrete block, and stone, available in a number of forms. Common examples of such structures which might be found around the home are garden walls and barbecues. This type of assembly can also be used in slab construction, for patios, walkways and drives, or in veneer construction where a facing of tile, brick or thin stone covers a subsurface of either wood frame construction or masonry structure.

FORM-AND-CONCRETE MASONRY

The third category involves the use of concrete mix poured into forms or molds, finished, and allowed to cure into a solid mass in the form of either a complete structure or a base to which a superstructure of wood frame or unit masonry construction will subsequently be attached. You can make heavy-section pours, as in a set of steps, for instance, or slab pours like walks or driveways. Small units can be made in the same manner, such as pavers or flagstones, decorative concrete screen blocks to be used in unit masonry construction, or individual items like flower boxes, patio furniture, posts and railings. Many do-it-yourselfers shy away from "wet" masonry work of this kind, under the mistaken impression that the procedures are highly specialized and difficult, beyond the limits of their skills. For some reason, there seems to be a general reluctance to dive in, and a fear that a tremendous amount of work might be involved which, because of a lack of expertise, would result in an unsatisfactory product. And yet, nothing could be farther from the truth. The fact is, "wet" masonry work is not all that difficult (save for the physical effort needed, which admittedly is often considerable), and the techniques can be readily learned.

There are only three things you need: knowledge, a bit of experience, and a few tools. It won't take you long to figure out how to put together "building-block" masonry projects, such as walkways, barbecues, walls, or planters. Nor is there any great difficulty involved in the poured or cast types of projects, such as garden pools, birdbaths, driveways and the like. The key is in knowing what to work with, and how to go about the job. To gain a complete understanding of "wet" masonry, the starting point is with *cement*.

CEMENT

Cement is where everything begins; it is the basic constituent that holds a masonry project together. By general definition, cement is a binding element or agent, a substance that makes objects stick together. Specifically, and with respect to masonry, cement is a dry powder composed of a mixture of *argillaceous* (derived from clays) and *calcareous* (derived from calcium carbonate/limestone) natural materials. Some of the individual ingredients of this mixture are alumina, silica, lime, magnesia, and iron oxides.

When the proper proportion of water is added to these conjoined materials, a dense, semi-solid mass is formed. Almost all of the water eventually evaporates from the mass, but the mass does not congeal by a process of "drying out" as we normally use that term. Instead, a chemical reaction known as *hydration* takes place between the cement and the water. Hydration begins soon after the water comes in contact with the cement, and as the elements of the material begin to combine, heat is given off. The mass slowly begins to "set up," the water gradually disappears, and the mass begins to harden (cure). The curing process goes on for a considerable period of time. This period varies depending upon the circumstances, but may extend to several weeks before full strength is realized.

Cement is not used by itself, because it lacks strength, durability, and other qualities critical to a construction material. To compensate for these deficiencies, an *aggregate* of one sort or another is added to the cement, thereby decreasing costs and greatly increasing effectiveness and usability. The cement serves as the glue to hold all the bits of aggregate together. This gives rise to four familiar building materials that we see around us every day: *concrete, mortar, grout,* and *stucco*. Each material has its own variations and uses (we'll delve into this subject a bit later). There is also a fifth material, *plaster*, which will not be discussed here since it is primarily an indoors, out-of-the-weather material not used in outdoor masonry projects.

Portland Cement

The term *portland cement* is frequently used interchangeably with the word cement, and has come to mean one and the same. Portland cement comes from neither Maine nor Oregon, but has

Type	Use Conditions	Compressive strength, % of Type I in days			
		1	**7**	**28**	**90**
I	General purpose	100	100	100	100
II	Moderate sulfate Low heat producing	75	85	90	100
III	"High-Early-Strength" Quick form removal Early service Early winter curing	190	120	110	100
IV	Slow curing Very low heat producing Large-mass pours	55	55	75	100
V	"Sulfate-Resistant" Severe sulfate action	65	75	85	100

Fig. 2-1. ASTM cement types.

become over the years a generic term stemming from a particular English manufacturing process which was developed in the early 1800s. When the manufacturer developed his cement, it seemed to him that the color of the cement was similar to that of the bedrock of the Isle of Portland, so he tacked the name onto his own product. Somehow, the name has stuck through all these years. Portland cement, whose gray color is familiar to all of us, is made from a carefully controlled and precisely proportioned combination of raw materials. After several steps of processing, these materials are fired in a kiln to make what are called *clinkers*. The clinkers are pulverized with a small amount of *gypsum* to make the final product.

ASTM Cement Types

The standards under which portland cement is made are set up by ASTM (American Society for Testing and Materials), and there are five distinct types available (Fig. 2-1).

The cement which most often fills the bill for home masonry projects, and the one that you will probably get from your local supplier unless you specify otherwise, is known as ASTM Type I. This is the all-around, general-purpose cement which is good for just about any application. ASTM Type III is known as *high-early-strength* cement because it is designed to cure earlier than other types. This cement is used primarily where for some reason or other the construction forms must be stripped away soon after the pour is made, or where fast cures in cold weather are necessary. The do-it-yourselfer should be able to plan his program so as to avoid having to use this type of cement, but it is available if needed.

ASTM Type II is designed to be used where the finished project will be in contact with either soil or water of a moderate sulfate content, and ASTM Type V will hold up where sulfate conditions are severe. If sulfate (a salt or ester of sulfuric acid) is a problem in your area, the fact will be well known and one or the other of these two types of cement will be available. The average do-it-yourselfer is not likely to be much interested in ASTM Type IV, which is a low-heat-generation type of cement used in making large monolithic pours.

Special Types of Cement

In addition to these general types of cement, there are a few special varieties. Two are of particular importance to the do-it-yourselfer. The first is called *white portland* cement, or *white porcelain* cement. The characteristics of this cement are the same as those of the gray types, except for the fact that the color is pure white because manganese and iron oxides have been kept to an absolute minimum during the manufacturing process. The importance of this cement lies strictly in its appearance; it is used where for architectural purposes the gray color is not wanted. Also, this is the cement which is used when a colored mortar or stucco is desired. Practically any earth-tone color can be produced using white portland cement as a base, but the same is not true of the gray.

The other special cement is called *masonry* cement. This material has a few extra ingredients added to increase workability and plasticity and to allow a high degree of water retention. Masonry cement is intended for mixing mortars where smoothness, workability, and a relatively long period of useful working time before set-up commences is important. Several other special varieties are available, but would only be of interest in exceptional circumstances that the homeowner is not likely to encounter.

Three of the five standard types of cements, I, II, and III, are also available in what is known as *air-entraining* cement. These are designated by tacking an A onto the type number. Special ingredients are added to the cement which cause a myriad of tiny air bubbles to be generated in the concrete. This is of particular importance to certain kinds of projects built in areas of severe winter weather. The air bubbles greatly improve the resistance of poured concrete to the repeated freeze-thaw action that occurs throughout the winter days. Furthermore, this type of concrete is much more resistant to the degrading action of snow- and ice-removal chemicals such as road salt. For those who are interested in building exposed patios, walkways, driveways and the like, the use of air-entraining cement may well be a worthwhile move.

Oh yes, one more point. Cement comes by the bag, or sack, containing exactly 1 cubic foot of material. Each sack weighs 94 pounds, so when you set out for the lumberyard, take a friend along.

WATER

Concrete is a recipe of four ingredients: cement, water, sand, and aggregate. We've already investigated cement; now let's look at water—the medium for blending the ingredients.

Not much to be said here. If you can drink the water, for the most part you can also use it to make concrete. The one exception to this is water which contains a high level of sulfate. This might be drinkable, but still it would not make for a good mix. If you are unsure as to the status of your own water, check with local health department officials; they'll know. On the opposite side of the coin, sea water, which is definitely not drinkable, can be used with a certain degree of success in mixing concrete. The quality of the concrete will not be as good as it could be, however, and salt water should not be used unless absolutely necessary.

SAND

Sand is actually an aggregate, but because it is classified in a different fashion, it is usually considered as a separate and individual part of the concrete mix. But sand is just—well, sand, isn't it? No, it isn't, at least when you're talking about concrete.

Sizes and Grades

Sand may also be called *fine aggregate*, with grains running to a maximum size of 1/4-inch diameter. Sand may be recovered from a natural granular deposit or it may be ground down from rocks and gravel. A given amount of sand will consist of particles of an almost infinite number of sizes. The usual procedure is to assign a grading number, called a *fineness modulus*, to a given batch of sand. Without going into the dreary details, this is done by sifting the sand through a series of sieves, each sieve of a different aperture size. Computations are made according to the amount of sand which passes through and the amount which remains behind. From this an FM is assigned. The lower the FM number, the finer the sand.

Fortunately, you don't have to be able to figure these FM numbers, as they all relate to three trade grades of sand. Fine sand FM numbers range from 2.20 to 2.60. Medium sand runs from 2.60 to 2.90, and coarse sand ranges from 2.90 to 3.20. The numbers are often important to engineers and professional concrete workers confronted with tricky or difficult specifications. Their importance to you is less critical, but still worthy of note.

As you might expect, the medium grade of sand is perfectly utilitarian and can be used in most concrete mixes. If the mix contains coarse sand, then the concrete will be quite plastic and workable and flow readily, all other things being equal. This can save some time and effort when you're pouring a relatively large job. Coarse sand is also used in a mix which will be finished by machine rather than by hand. A fine grade of sand, however, works best where the finishing is to be done by hand, because a smoother and better-looking finish can more easily be obtained. This type of mix is also easier to work, and with better results, where small batches are used in small projects and where a smooth and uniform surface is desired.

Another factor involved is the richness or leaness of the mix. A *rich* mix has a proportionately large amount of cement, and consequently a coarser grade of sand can be used in the interest of both efficiency and economy of pouring. A *lean* mixture, on the other hand, contains a relatively small amount of cement, and the finer grades of sand are used so that the particle bonding will be totally effective. Also, when small-size aggregates are used in the mix, a fine grade of sand is normally used in conjunction with them.

It is also worth noting that in no case should the sand be of uniform size. That is, whatever the grade of sand is, the grains should range from the top grade size down to tiny particles. Uniform sizing of the particles weakens the concrete, while a complete range of particle sizes allows maximum bonding and consequent strength.

The sand particles should also be both hard and sharp. The relative hardness of the grains plays a part in the ultimate hardness and strength of the concrete. Sharp-edged and slightly irregular grains also contribute to the formation of a stronger finished product.

Can you use just any old sand for mixing concrete? Sometimes you can, but not always. If you have a ready source of sand from a local bank, this would be well worth checking, especially if the material is free. You can screen the sand yourself, using a homemade screen. Build a frame from two-by-fours and cover the bottom with screening which has a mesh of an appropriate size. Use a 1/4-inch mesh size for coarse sand, a 1/8-inch size for medium sand, and a 1/16-inch mesh for fine sand. This will not necessarily give you the same results as are obtained when sand is run through commercial sieves, but should give you a maximum grain size, as well as a reasonable range of particle sizes, to make a perfectly usable concrete mix.

Clean and Dirty Sand

One thing you must watch out for, though, is the cleanness of the sand. Dirty sand does not make a good concrete mix because foreign substances interfere with the proper bonding of the cement-water mix to the aggregate particles. The more dirt in the sand, the weaker will be the finished concrete mix. Strength can be seriously impaired and the durability and weatherability markedly reduced. In natural sandbanks you will frequently find lumps of clay or loam, bits of root or other organic material, quantities of dust-like silt and perhaps other odds and ends as well. You can make a quick check by rubbing a small amount of undisturbed subsurface sand around on the palm of your hand. If your hand is dirty after you release the sand, the chances are that this sand is not suitable for a decent concrete mix.

This simple test, however, is really a bit too simple, because it does not reveal the presence of a thin clay coating over the sand particles, which is commonly found in natural sandbanks. A better test is to remove a small amount of sand from an undisturbed area and let it dry. Put about 2 inches of the dried sand in the bottom of a quart jar, and then fill the jar about three-quarters full with water. Shake the jar vigorously and let the mixture stand for at least an hour. If more than a 1/8-inch layer of sediment settles out on top of the sand, the sand is unusable. Anything less than that amount is all right.

Large amounts of organic material in the sand can also destroy its effectiveness, at least to a degree. If you suspect that this might be a problem, you can make another test with caustic soda to find the proportion of organic material present. Use a solution of 1 ounce of sodium hydroxide dissolved in 1 quart of distilled water. Place 4 1/2 ounces of the sand to be tested in a 12-ounce bottle and add enough of the solution to bring the level up to 7 ounces. Cork the bottle with a rubber stopper, shake vigorously, and let stand for at least 24 hours. If the liquid remains clear, no organic material is present. A light yellow or straw color indicates the presence of some organic material, but not enough to create trouble. If the liquid is tea-colored, the sand is not usable.

Though dirty sand should not be used in a concrete mix, if you have no alternate source of supply you can wash the sand yourself. This can be done by using a garden hose to direct a strong stream of water into a large container which can be continuously drained without losing the sand. If the sand is heaped in large piles, a reasonable job of washing can be accomplished by turning the hose on the pile and shifting the bulk material around from time to time. Let most of the moisture drain away from the sand; wet sand does not handle well at all. Incidentally, don't use beach or alkali sand unless it has been thoroughly washed to remove salts and impurities.

The easiest way to avoid any problems of this sort is to purchase your sand from a reputable gravel pit operator who supplies these materials on a regular basis to masons and ready-mix concrete companies. Then you can be reasonably well assured of getting the right kind of material of good quality. Be prepared, however, to pay for that service.

AGGREGATES

The aggregate in a concrete mix is properly called the *coarse aggregate* to differentiate it from the sand, which is the fine aggregate. Coarse aggregate is the material which makes up the bulk of the volume of most concrete mixes. Coarse aggregate is also graded as to size, though the system used is simpler and easier to see than the one for sand. The smallest coarse aggregate takes up where sand leaves off, at the 1/4-inch size, and ranges upward to a maximum of about 3 inches in diameter. For the most part, sizes larger than 1 1/2-inch diameter are seldom used, especially for projects such as might be found around the home. In fact, for the sort of projects we will be discussing later, 3/4-inch diameter is usually the maximum size recommended.

Desirable Qualities

Where available, natural gravel is the preferred material for aggregate. In this context, gravel is usually taken to mean pebbles and small rocks or pieces of stone in a wide range of sizes, with a good mix of all the major size levels measured in fractions of an inch. Where gravel is not available, crushed stone is the usual alternative. This consists of larger rocks which are run through a crusher and broken up into various smaller sizes. In either case, the aggregate is screened to arrive at the proper maximum size level. As with sand, the aggregate pebbles should be hard and tough, and relatively sharp-edged. They also should be essentially spheroidal, rather than grossly egg-shaped or in long splinters or slivers. There should be a good even mix of sizes from the largest allowable for a given concrete mix to the smallest size (Fig. 2-2). This ensures maximum strength in the cured concrete, but is also necessary from a purely practical standpoint. A uniform mix of aggregate in one size, with or without the inclusion of odd-shaped pieces, means that the aggregate will not pack together well and there will be a substantial amount of air space left. That air space must then be filled with the cement/water paste, which increases the cost per unit volume of the concrete mix by a surprising amount. With a full range of aggregate sizes, the smaller pieces fill the voids between the larger pieces, and less cement is necessary.

Fig. 2-2. Well-graded gravel aggregate.

Cleanness of the aggregate is just as important as cleanness of the sand. There should be no lumps of clay or loam, pieces of roots or branches, or other organic material, and no rubbish such as broken glass or pieces of rusty metal. The rock should also be free from any clay or silt coating which could interfere with the proper bonding action of the cement. The best bet is to use washed gravel, which will be almost entirely free from any small particles. If only unwashed gravel is available, you can wash it yourself by watering the pile of gravel liberally with a garden hose, then shoveling off the top layers as needed, and continuing the washing as the pile diminishes. Unwashed aggregate is frequently used in concrete mixes, but should not be unless it is obviously fairly clean; you can tell by visually inspecting a few random shovelfuls.

Lightweight Aggregates

Lightweight aggregates are just that—aggregates which are considerably lighter in weight than rock, and result in a concrete mix which is lighter in weight than normal. Whereas a standard concrete mix weighs in at 145 pounds per cubic foot, lightweight aggregate concrete may run anywhere from 65 to 125 pounds per cubic foot. Some of these lightweight aggregate materials are scoria or slag, cinders or pumice, perlite, and micaceous minerals, all of which have been further expanded by extreme heat. Lightweight aggregate concrete is a rather specialized type of concrete which is not normally found around the home, except perhaps as a fireproof lining in woodstoves and fireplaces, and so will not be dwelt upon here. However, if you have an application for which you feel that a concrete of moderate or extreme light weight would be better than the standard variety, the subject would be worth looking into further.

ADDITIVES

Most of the time there is no need to add anything to the concrete mix, as the standard ingredients will do the job nicely. There are occasions, however, when a little something extra is called for and can make the job easier and more effective.

Air Entrainment

One of the most worthwhile additives, or admixtures as they are often called, is an air-entraining powder. As noted earlier, cement may be purchased with this ingredient already included, but it can also be added to a standard type of cement at mixing time. The resulting concrete has the same advantages afforded by an A-Type cement mix: increased durability under freeze-thaw action, and resistance to the deleterious effects of snow-removal chemicals. An added benefit is that the concrete mix is more plastic and workable, which means that it pours better, places more readily, and is more easily finished.

Plasticizers

Where there is reason for a concrete mix to be particularly plastic, as for easier placement or workability, then a *plasticizer* is added to the mix. Mortar, for instance, must be very plastic and easy to work, and about the most common plasticizer used for this purpose is lime. Just a small amount of lime makes a batch of mortar easier to work with and will have the same effect on concrete.

Another commonly used plasticizer is simply a little extra cement. A small amount added at the dry-mix stage improves the workability of the concrete, or the mortar, though in the latter case lime is more effective. Similarly, when the dry-mix batch is made up, you can use a finer grade of sand than might ordinarily be called for, thereby achieving a more plastic and workable mix. This mix also makes for an easier hand-finishing job, resulting in an exceptionally smooth surface if the proper care is taken. A concrete mix which contains an air-entraining ingredient also will be more plastic than a mix which does not. In lieu of another plasticizer, an air-entraining ingredient can be added to a mix in the interest of better workability, whether the air-entraining characteristics are necessary or not. Certainly, no harm can come from using this admixture.

Retarders and Accelerators

Temperature and weather extremes play an important part in the manner in which concrete cures. High temperatures cause too rapid hydration, while low temperatures slow down the curing process, and if low enough can present the danger of freezing. Strong winds, especially when coupled with low humidity, can foul up the curing process, and so can rainfall. Professional concrete workers frequently run into these difficulties because they must keep to a certain work schedule despite the weather. As a do-it-yourselfer, however, you should be able to arrange your own schedule to avoid weather conditions which are not amenable to this type of work.

But if you do get stuck, there are both additives and work procedures which will help to alleviate the difficulties. As far as the additives are concerned, there are two principal ones. The first is called a retarding admixture, and is designed to hold back the hydration process, allowing the concrete to cure at a slower pace in hot weather. A material called *pozzolan* is often used for this purpose, and is added to the dry-mix to slow the heat release of the concrete.

The second admixture is calcium chloride; it is often used to accelerate the curing time of the concrete in cold weather. It's used when the temperatures get down around the 40-degree mark, but will not serve as an antifreeze in the mix to allow pouring in below-freezing temperatures. Though salt and concrete do not normally get along well together, in this case the concentration is so small that no harm is done. A 2 percent solution is most generally used, with the calcium chloride added directly to the water used for mixing the batch. It cannot be added to the dry-mix in crystal form, as it will not properly dissolve.

Neither the accelerating nor the retarding admixture should be considered as an easy way out so that you can pour concrete under dubious conditions. Rather, they should be avoided whenever possible, as they increase both the expense and the effort required, and result in a job which may be of lower quality than would be possible under good conditions. Other procedures, which are more effective when you have no choice but to go ahead and pour, will be discussed a bit later on.

Water-Reducing Admixture

There is one more additive which deserves mention. The more water you add to a given quantity of dry-mix concrete, the lower will be its strength when finally cured. The less water you use, the stronger the concrete will be. The object, then, is to use as little water as possible in the mix. But still, there must be enough water in the mixture to permit decent workability and so that the wet concrete can be easily placed. Where the maximum possible strength is desired, you can use an additive called a *water-reducing admixture*. By including this ingredient, you can reduce the amount of water needed for a workable mix, and thereby increase the cured strength. For many projects, especially those found around the home, this additive is not needed. On the other hand, there may be occasions when you will find it quite helpful.

PROPORTIONS

Mixing up a batch of concrete is a lot like making a cake. If you use the proper ingredients in the right proportions, you will come out with an excellent finished product. But if any part of the formula is incorrectly followed, all you will have is a wasted batch of materials. Concrete is mixed in two steps: first the dry-mix, and then the wet mix. The proportions of the ingredients used in the dry-mix stage are important. The proportion of water to dry-mix is critical. In both cases, the specific proportions are somewhat dependent upon the nature of the job being done.

Determining the Proper Proportion

There are three general methods for determining just what the proportions of a particular mix should be. The first and most commonly used method is to consult existing charts and tables. Some of these are quite complex and take a great many variable factors into account, while others are so simple as to be almost elementary. Even the simplest of tables, however, will work quite nicely for home masonry projects under most circumstances. The second method involves what is known as a *slump test*, which we will investigate a bit later. The third method usually makes some use of the first two methods, but is dependent upon a large amount of practical field experience to arrive at on-the-spot decisions, in light of the particular job demand, working conditions, and the materials at hand.

Dry-Mix Ratios

Dry-mix ratios are usually expressed in terms of cubic feet or cubic yards. The first figure in the ratio series refers to the cement (1 sack of cement is equal to 1 cubic foot). The second number refers to the sand, and the third number refers to the aggregate. Thus, a 1:2:3 ratio means that the mix consists of 1 cubic foot of cement, 2 cubic feet of sand, and 3 cubic feet of aggregate. Similarly, a 1:3:5 ratio would mean that the mix is in the proportion of 1 cubic foot of .cement to 3 cubic feet of sand to 5 cubic feet of aggregate. Note that these ratios do not have to denote quantities, but only proportions. The 1:2:3 dry-mix can also be read as 1 part of cement to 2 parts of sand to 3 parts of aggregate. Each unit part could just as easily amount to 10 cubic yards, or any other number or measure, as 1 cubic foot.

To further confuse the issue, note that you cannot determine the volume of concrete needed for a given project by following the dry-mix ratio figures. For instance, the 1:2:3 mix ratio, if expressed in cubic feet, would appear to equal a total of 6 cubic feet of finished concrete. But this is not so—there is a considerable shrinkage factor. More about the business of figuring quantities will be discussed later.

As well as being classed in cement-sand-aggregate ratios, dry-mixes are also categorized according to the amount of cement included relative to the sand and aggregate. There are four classes of mix: rich, standard, medium, and lean.

The *rich* mixture is in a ratio of 1:2:3. This makes a waterproof cement and one which is readily workable. This is a good mix for garden pools or fountains, driveways, or walks. It also works well for any sections which are to be 4 inches to 8 inches thick, and for fenceposts, planters, and the like.

The *standard* mixture consists of a 1:2:4 mix. Though the cement and sand proportions remain the same as in the rich mixture, the amount of coarse aggregate is doubled. This is a good workable mix for general-purpose projects of medium size, or with reinforcing, for larger sizes. Appropriate applications might be for garage floors, columns and arches, and for reinforced sections from 8 inches to 12 inches thick.

The *medium* mixture is made up of 1 part cement, 2 1/2 parts sand, and 5 parts aggregate. This ratio is normally used for heavier work, such as footings, foundations, basement walls, piers, and the like.

The *lean* mixture is mixed in a 1:3:6 ratio, and is principally used for massive concrete work such as the foundations for large buildings.

The mixtures listed above are more or less standard, but there are many other ratios in common use as well (Fig. 2-3). For instance, a 1:2:2 ratio works well for making up thin sections of concrete from 2 inches to 4 inches thick, and is good for patio furniture, benches, fenceposts, and similar relatively small articles which will be cast and then set in place—a birdbath, for example. For projects around the home, mixtures up to about 1:2 1/2:3 1/2 or 1:2:4 are perfectly usable. Anything much heavier than this would likely be unnecessary and would only add to the construction problems.

Keep in mind that by and large the richest mixtures are the easiest to work with, they result in tough and strong finished projects, and sometimes require no reinforcing. They are also the most expensive, though when used in small quantities the difference in cost is not that dismaying. The leaner mixtures are harder to work, but less expensive. In some projects, a large quantity of coarse aggregate relative to the other materials makes the concrete more difficult to finish, especially in small projects or thin sections.

Mix Ratio	Section Thickness	Use
1:1 1/2:3	2″-4′	Furniture, posts,
1:2:2		tanks, planters.
		Corner posts, tanks,
		planters, paving.
1:2:3: (Rich)	4″-8″	Extremely watertight.
		Abrasion resistant.
		General purpose.
1:2:4 (Standard)		Walls, steps, paving.
1:2 1/2:3 1/4	4″-8″	Watertight work.
		Reinforced work.
		Footings, walls,
1:2 1/2:5 (Medium)		foundations.
		Footings, foundations.
1:3:5	Mass	Massive pours.
1:3:6 (Lean)	Mass	Not watertight.
		Moderate strength.

Fig. 2-3. Common concrete mix ratios and their applications.

CHOOSING THE DRY-MIX

The first step in choosing a dry-concrete mix for whatever project you have in mind is to determine the size of the fine aggregate, or sand. The usual choice is a medium sand, tending toward the finer grade if hand-finishing is to be done, or where the aggregates will also be sized rather small, or where the mixture is to be a lean one from a cement standpoint. Coarse sand is often used where ease of placement is important, or where finishing will be done by machine. Since most home projects are relatively small and the finishing work is done by hand, a fine to medium grade of sand is about right. If the sections are to be particularly thin or if small articles are to be cast, use a fine sand.

Size of the Coarse Aggregate

Then you must choose your coarse aggregate size. There are several considerations here which must be taken into account. The minimum size of the aggregate should be 1/4-inch diameter. The maximum size is dependent upon several factors. Generally speaking, if the thickness of the section is 4 inches or less, or if the section is somewhat thicker but is reinforced, then the aggregate can be graded up to 1-inch diameter. A commonly used size which works well in smaller home projects is the 3/4-inch diameter, a size readily available at any gravel pit. In any case, the maximum size of the aggregate should be no greater than 1/5 of the thickness of the concrete section. The exception is with slabs which are poured on the ground, where the aggregate can be as large as 1/3 the thickness of the slab. Concrete with large-size aggregate is harder to work with but is more economical. On the other hand, smaller sizes of aggregate generally require the use of more water, with a resultant loss in strength. The idea is to compromise to whatever extent is necessary, but to end up using the largest maximum aggregate size that seems reasonable and prudent for a given task.

The Proper Ratio

Finally, determine the most appropriate dry-mix material ratio for your intended project. It's easiest to stick with round numbers here, especially when you are using tables to determine the final concrete mix. The finest usable concrete mix is generally considered to be a 1:1 1/2:3 or a 1:2:2 ratio, ranging upward to perhaps as high as 1:2 1/2:5 for large home projects. Most of the time you will find that a 1:2 1/2:3 1/2 or a 1:2:4 ratio will be fully satisfactory at the upper end of the scale. This is the time, too, to decide whether or not you need to include any additives or admixtures.

FIGURING QUANTITIES

Once you have decided on the details of what the dry-mix is to be, the next step is to figure out the quantities required for the project. The volume of wet-mix concrete that you will need for your project is not the same as the volume of the dry-mix. This is because each of the individual materials which go into the dry-mix contains a certain amount of air. When the ingredients are mixed together dry, the fine aggregates lodge in between and fill up the spaces around the larger aggregates, and the cement powder further fills the spaces around all of the particles. When water is finally added to the mix, nearly all the air is displaced and the voids completely disappear, resulting in shrinkage. This is the reason for using a wide range of gradation in both sand particles and in coarse aggregate, so that there will be a continuum of particles available to fill all of the various sizes of voids in the mix and make a dense, solid concrete.

Shrinkage and Overrun

The shrinkage is a variable factor, depending upon exactly which aggregates are used. The easiest way to do your figuring is to use tables prepared for the purpose. Keep in mind, however, that these tables are seldom absolute, and may cause you to err by 10 percent one way or the other. It's a good idea to mix up a bit more concrete than the figures call for. It's also a good idea to have a couple of small projects on hand, like a precast fencepost, wood posts to be set in concrete, or something of that sort, so that if you have concrete left over from your principal project, you can utilize the excess and have no waste.

How Much Dry Material?

There are three basic approaches to determining how much dry material you will need to make up a given volume of concrete. The first is to obtain the volume yield of concrete by means of the dry-mix proportions, as shown in Fig. 2-4, and calculate from there. The second is to consult a table such as the one in Fig. 2-5, which lists the quantities of materials needed to make up one cubic yard of concrete. You then relate this to the necessary volume for your project. The third method is more precise and more difficult to use, and resorts to tables which take into account the fineness modulus of the sand and the size of the aggregate, are based upon the aggregates by weight, and read out in cubic-foot yield per sack of cement.

The first step in getting a set of exact numbers for your project is to determine the volume of concrete that you will need. Suppose, for instance, that you are going to build a patio. The patio is 20 feet wide and 30 feet long. You can determine the volume in terms of

Cement	Sand	Gravel	Yield
1	1.5	3.0	3.5
1	2.0	2.0	3.4
1	2.0	3.0	3.9
1	2.0	4.0	4.5
1	2.5	3.5	4.6
1	2.5	4.0	4.8
1	2.5	5.0	5.4
1	3.0	5.0	5.8
1	3.0	6.0	6.4

Fig. 2-4. Cubic-foot (or unit) yield of various concrete mix ratios.

either cubic feet or cubic yards, but unless the project is quite small, cubic yards is the easiest way to go. To find the volume you must multiply the length times the width times the thickness (or height). The area, then, is 600 square feet (length times width). Earlier you

Mix	Cement (Sacks)	Sand (Cu. yd.)	Gravel (Cu. yd.)
1:1 1/2:0	15.5	0.86	----
1:2:0	12.8	0.95	----
1:2 1/2:0	11.0	1.02	----
1:3:0	9.6	1.07	----
1:1 1/2:3	7.6	0.42	0.85
1:2:2	8.2	0.60	0.60
1:2:3	7.0	0.52	0.78
1:2:4	6.0	0.44	0.89
1:2 1/2:3 1/2	5.9	0.55	0.77
1:2 1/2:4	5.6	0.52	0.83
1:2 1/2:5	5.0	0.46	0.92
1:3:5	4.6	0.51	0.85
1:3:6	4.2	0.47	0.94

Fig. 2-5. Quantities of materials needed to make 1 cubic yard of concrete with various mix ratios.

had settled on a suitable thickness for the patio of 4 inches, or a third of a foot. Multiplying 600 square feet by 1/3 gives you a total of 200 cubic feet. Dividing this by 27 (the number of cubic feet in a cubic yard) shows that you need approximately 7.4 cubic yards of concrete to do the job (Fig. 2-6).

To take another example, suppose that you plan to build a low wall with a footing underneath. Here you would figure the two sections separately, the footing by itself, and the wall section by itself. If the footing is 2 feet wide, 1 foot thick, and 10 feet long, its volume would be 20 cubic feet. If the wall which will sit atop the footing is 1 foot thick, 2.5 feet high, and 10 feet long, its volume will be 25 cubic feet. The total, then, is 45 cubic feet or approximately 1.7 cubic yards of concrete needed (Fig. 2-7).

Calculating for Irregular Structures

If the projects are irregular in shape or general configuration, then the best approach is to divide the project up into bits and pieces and figure each separately to arrive at a total. For instance, a patio might be triangular at one end, curved at the opposite end, and rectangular in the middle. In this case, you would figure the volume

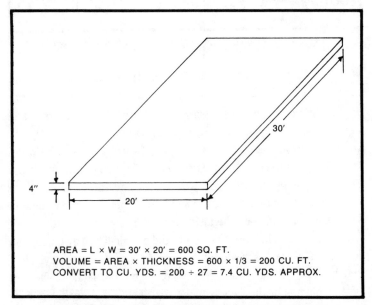

AREA = L × W = 30′ × 20′ = 600 SQ. FT.
VOLUME = AREA × THICKNESS = 600 × 1/3 = 200 CU. FT.
CONVERT TO CU. YDS. = 200 ÷ 27 = 7.4 CU. YDS. APPROX.

Fig. 2-6. Figuring concrete requirements for rectangular patio.

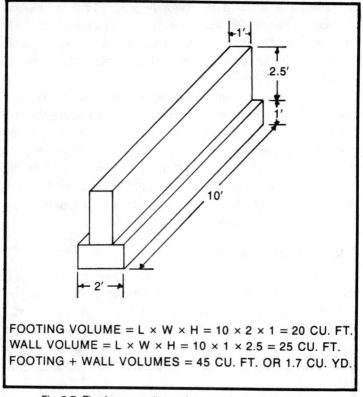

FOOTING VOLUME = L × W × H = 10 × 2 × 1 = 20 CU. FT.
WALL VOLUME = L × W × H = 10 × 1 × 2.5 = 25 CU. FT.
FOOTING + WALL VOLUMES = 45 CU. FT. OR 1.7 CU. YD.

Fig. 2-7. Figuring concrete requirements for footing and wall.

of the rectangle, then the volume of the triangular end piece, and then the volume of the curved end piece, and add them all together.

If the project were a square-walled planter in the form of an open-topped box, you could figure each wall section separately and then the bottom section, by multiplying the height times the width times the wall thickness for each piece. Then add them up. If the planter were cylindrical, you would first find the volume of the cylinder based upon the outside diameter. Then find the volume of the cylinder based upon the inside diameter, and subtract the inner from the outer. The bottom piece would be calculated by finding the area of the circle described by the outside diameter of the cylinder, and then multiplying by the thickness of the bottom section. The sum of the volumes of the two sections represents the amount of concrete the structure will require (Fig. 2-8).

Multisided structures or odd shapes are much more difficult to figure. Small garden pools, for instance, which are dish-shaped or irregular like half a peanut shell are not easy, and as often as not you will have to make some educated guesses as to just how much material you will need. Conical, pyramidal, and other more or less regular geometric shapes can usually be calculated with the use of the appropriate mathematical formulae, and by breaking the structure down into its component simple shapes whenever possible.

Volume of Raw Material Versus Finished Volume

Once you come up with a reasonably accurate calculation of the quantity of in-place concrete required, you can proceed to relate this to the amount of raw material that you will need to mix the batch. By way of example, let's assume that you have determined that the footing for your decorative garden screen wall will require 2.8 cubic yards of concrete. Since footings don't require an exceptionally

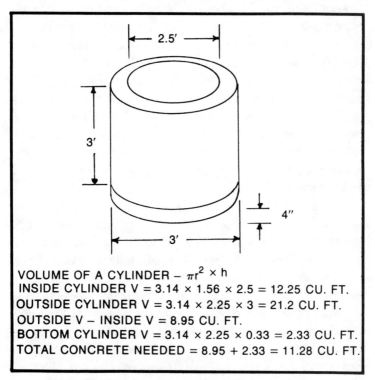

VOLUME OF A CYLINDER – $\pi r^2 \times h$
INSIDE CYLINDER V = 3.14 × 1.56 × 2.5 = 12.25 CU. FT.
OUTSIDE CYLINDER V = 3.14 × 2.25 × 3 = 21.2 CU. FT.
OUTSIDE V – INSIDE V = 8.95 CU. FT.
BOTTOM CYLINDER V = 3.14 × 2.25 × 0.33 = 2.33 CU. FT.
TOTAL CONCRETE NEEDED = 8.95 + 2.33 = 11.28 CU. FT.

Fig. 2-8. Figuring concrete requirements for cylindrical planter.

high-strength concrete and don't have to be watertight, you have decided that a 1:3:5 dry-mix ratio will do the job. On consulting the table of materials required for making up 1 cubic yard of concrete, you see that the 1:3:5 line of the table calls for 4.6 sacks of cement, 0.51 cubic yards of sand, and 0.85 cubic yards of aggregate. Since you will need 2.8 cubic yards altogether, you multiply all of those figures by 2.8. This means that you will need approximately 12.9 sacks of cement (about 0.5 cubic yards), approximately 1.4 cubic yards of sand, and approximately 2.4 cubic yards of aggregate. Observe that this totals about 4.3 cubic yards of material to make up your needed 2.8 cubic yards of in-place concrete. Quite a shrinkage factor!

Playing Safe by Adding a Little Extra

Now, on a job like this, you don't want to run the risk of coming up short of concrete, since the footing is best made all at once, or in a consecutive series of batches. Running short unexpectedly means extra problems and extra work, not to mention a finished job that may be somewhat less than satisfactory. Since concrete mix tables can vary as much as 10 percent, it does no harm to figure in a little extra raw material in the event the formula falls short. Add 5 percent, or even 10 percent if you wish, to your figures. Of course, if the raw materials run long, then you could end up with as much as 20 percent left over. But unbroken sacks of cement can usually be returned to the supplier, and the extra aggregate and sand you can probably turn to some other use. Or, keep the whole works on hand in dry storage for the next project.

Another point to remember is that in most concrete pouring jobs there is inevitably some amount of waste. There seems to be no way to avoid this. It is reasonable to expect waste amounting to a minimum of 5 percent of the total in-place volume. On larger projects, waste may amount to 10 percent. On small projects the quantity of waste isn't significant, but on larger jobs the quantity of concrete that simply vanishes can be considerable. This occurs often enough that you can run short if you don't make prior allowances. On jobs of up to a cubic yard or so, it's a good idea to add 5 percent to the total amount of concrete needed. On the larger jobs, add in an extra 10 percent. As mentioned earlier, you can have some small projects lined up and ready to go so that you can utilize the leftover mix if there is any.

Calculating Quantities for the Smaller Jobs

Many of the projects around the house are small and don't require a great deal of concrete. These jobs are usually figured in terms of cubic feet rather than cubic yards, but you arrive at the results in the same way. Let's say that you have designed a concrete bench to sit at the edge of your terrace. You have determined that you will need 6.5 cubic feet of mix to complete the job. Since the design of the bench is irregular and the volume difficult to calculate precisely, you decide to add an extra 10 percent for good measure, plus 5 percent for waste. This gives you a total requirement of approximately 7.5 cubic feet needed. The mix will be in a 1:2:2 ratio, using fine sand and small aggregate for ease of placement in the forms and so that you can sculpt and hand-finish the top edges and back portion of the bench easily and with good results. The 1:2:2 line in the table in Fig. 2-5 shows that you will need 8.2 sacks of cement, 0.60 cubic yards of sand, and 0.60 cubic yards of gravel to make the mix for 1 yard (cubic yards are generally shortened to just yards in speaking of concrete). Multiply these figures out as though you were working with cubic yards instead of cubic feet, using your figure of 7.5. This would give you 61.5 for the cement, 4.5 for the sand, and 4.5 for the gravel. Now, since there are 27 cubic feet in a cubic yard, you must divide all the numbers by 27. This would give you approximately 2.28 sacks of cement. The conversion factor from cubic

	Cement Cu. Ft.	Sand Cu. Yd.	Gravel Cu. Yd.	
Example: 7.5 cu. ft. of 1:2:2 mix needed.				
Materials -				
	8.2	0.60	0.60	-for 1 cu. yd.
Mult. × 7.5 -	61.5	4.5	4.5	-for 7.5 cu. yd.
Div. by 27 and convert to cu. ft.	2.28	4.5	4.5	-for 7.5 cu. ft.

Fig. 2-9. First method of determining required concrete amount.

Fig. 2-10. Second method of determining required concrete amount.

yards to cubic feet is 27, so you can save a step, and just change the unit measure of the last two figures to cubic feet instead of yards. This will give you 4.5 cubic feet each of sand and aggregate (Fig. 2-9).

Another method of making the calculation is to reduce the proportion figures to cubic feet before completing the formula. This is done by dividing all the cubic yard figures by 27, resulting in requirements for 1 cubic foot of concrete of 0.304 sacks of cement, 0.023 cubic yards of sand, and 0.023 cubic yards of aggregate. But again, the 27 factor comes in, so just change the last two figures to cubic feet so you don't have to reconvert. Then multiply by your needed quantity, 7.5 cubic feet, to arrive at 2.28 sacks of cement, 4.5 cubic feet of sand, and the same for aggregate (Fig. 2-10).

Example: 7.5 cu. ft. of 1:2:2 mix needed.

Materials -	Cement Cu. Ft.	Sand Cu. Yd.	Gravel Cu. Yd.	
	8.2	0.60	0.60	-for 1 cu. yd.
Convert requirement to cu. yd. - 7.5 ÷ 27 = 0.28 cu. yd.				
Mult. by 0.28	2.3	0.168	0.168	-for 7.5 cu. ft.
Mult. by 27, convert to	2.3	4.5	4.5	-for 7.5 cu. ft.

Fig. 2-11. Third method of determining required concrete amount.

A third method reduces your needed amount of concrete (7.5 cubic feet) to a decimal fraction of a cubic yard. Convert the cubic yard to cubic feet, 27, and then divide this figure into your required 7.5 cubic feet. This shows that you need 0.28 cubic yards of concrete. Now go back to the 1:2:2 ratio line and multiply the quantity figures by 0.28. This will result in a requirement of 2.3 sacks of cement, 0.168 cubic yards of sand, and the same for the aggregate. Then convert the cubic yards of sand and aggregate to cubic feet by multiplying by 27, for an answer of 4.5 cubic feet each (Fig. 2-11). The slight variations in the figures from method to method result from rounding off some rather long fractions. In all cases, a pocket calculator is a mighty hand tool to have around.

Preparing to Pour Concrete

As often as not, getting ready to pour concrete is much more of a task than the actual pouring and finishing operations. The most involved jobs consist of preparing the site and building forms to receive the concrete, and the excellence of the finished project will rest to a great degree upon how well this is done. Depending upon the nature of the project, there may be other considerations as well, such as the installation of reinforcement, keying separate pours together, tying a superstructure to the finished pour, or arranging for joints and grooves. All of this work must be either finished or planned for and laid out before the concrete mixing begins.

FORMS

Freshly mixed concrete is a semiliquid mass—a slurry—whose consistency ranges between that of a thick soup and a stiff porridge. Except in a few special instances where small amounts of rich mix are made up with so little water that the mix can be sculpted almost like a modeling clay, fresh concrete must be set in some sort of retaining structure or container. There it remains until it reaches sufficient strength to allow the supporting structure to be stripped away. Such supporting structures, regardless of their particular construction, are called *forms*. Certain types of forms may be called *molds*.

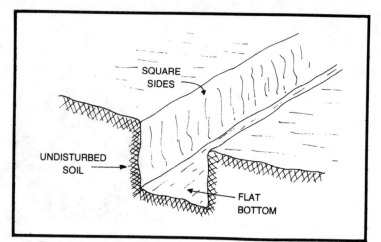

Fig. 3-1. Trench dug into undisturbed soil serves as footing form.

Earth Forms

In many cases, the earth itself can serve as the form. This is the easiest and most efficient way to contain the concrete, provided the situation lends itself to this process. Practically any concrete which is to be poured so that its surface is even with or below the level of the soil can be poured directly into an earthen form, provided that the sidewalls are of undisturbed native soil (rather than compacted or loose fill) and provided that the earth can be properly shaped and that it is stable enough not to fracture and collapse.

For instance, in making a footing for a garden wall, all you have to do is dig a square-sided, flat-bottomed trench to the dimensions of the footing, and pour the concrete into it (Fig. 3-1). Flat concrete work such as patios, driveways, or walkways use the earth for bottom support while the sides are held in place with narrow form boards. In some circumstances, as with the construction of a narrow walkway whose top surface will be level with the surrounding grade, it is possible to eliminate the use of side forms by excavating the earth in accordance with the intended size and shape of the walkway (Fig. 3-2). In all cases, considerable care must be taken when finishing the top surface of such concrete pours to avoid knocking debris and chunks of dirt into the fresh cement.

Posts, poles, piers and similar structures can be set in concrete which is held in place only by the earth surrounding it (Fig. 3-3). In

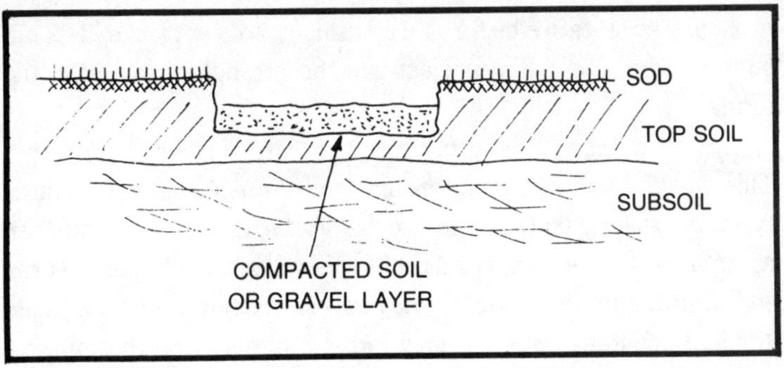

SOD

TOP SOIL

SUBSOIL

COMPACTED SOIL
OR GRAVEL LAYER

Fig. 3-2. Walkway or similar slab can be poured into shallow excavation in soil or sod.

fact, in the case of fence posts, which must be firmly anchored in the ground for sturdiness and stability, the concrete surrounding the post *should* be in direct contact with the earth to obtain the best support possible. Encasing a fencepost in concrete poured into a wood box form, and then setting the whole business in a hole and tamping earth around it, is really not much different than just digging a hole and planting the fencepost itself. Practically nothing is gained by the addition of the concrete, and it costs a considerable amount of labor.

Wood Forms

Wood is the most commonly used material for making forms, since it is readily available, easy to work with, relatively inexpensive, and can be dismantled for building new forms or converting to other purposes. When a project such as the base or platform for a barbecue is built in place, the bottom of the slab will rest upon the

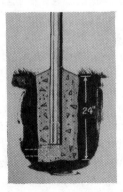

Fig. 3-3. Posts and poles can be set in concrete poured directly into a hole dug in undisturbed native soil.

ground. The sides are be formed up using boards or planks, depending upon the size of the project and the strength required in the forms.

Some items may be precast and set into position after curing. This would be true of concrete flagstone, for instance, or similar types of building units which are fabricated unit by unit and then assembled into the completed project. Some smaller items like urns and planters are designed so they can be moved if ever the need arises. Such items must be fully formed, at sides *and* bottom.

More complicated projects may require two or even several sets of forms employed in successive stages. The first section of the project is formed, poured and keyed to the next stage. When this operation is complete and the concrete is cured, the second set of forms is placed in position and the second pour made, and so forth.

The kind of wood that you use in building forms is dependent to some degree upon the nature of the project. For concrete work which will be hidden from view, or later covered with another surface such as stone veneer or stucco, you can use just about any old wood that happens to be handy. If the surface of the work will be exposed, however, and must be attractive to the eye, the form material should be perfectly smooth and have tight-fitting joints. Be advised that any imperfections appearing on the interior surfaces of the forms will be replicated on the concrete surfaces.

This situation can also work *for* you, rather than against. By properly arranging the inside surfaces of the forms, you can produce a pattern, either imprinted or in bas-relief, on the finished project. For instance, by using Douglas fir plywood, you can give the surface of the finished concrete a plywood grain effect. A sandblasted plywood will result in a driftwood finish. Or, you might start with a perfectly smooth form surface, and then attach a design made from pieces of wood molding nailed to the inner surface of the form. This will result in the identical pattern being impressed into the finished concrete.

The particular kind of wood that you use for forms and form supports really does not matter all that much. Construction-grade spruce is one of the best materials and is relatively inexpensive and readily available. White pine works well, but is more expensive and less durable. Plywood is frequently used, generally in a minimum

3/4-inch thickness. For structures of larger mass, 1-inch and 1 1/4-inch thicknesses are used. A special plywood is available which is specifically designed for form work, and is often used by professionals. This material, however, is quite expensive since it is intended for repeated hard usage; the homeowner would probably seldom have either the inclination or the need to purchase commercial form equipment. Professionals also often use specially designed metal forms which give a perfectly smooth finish to the surface of the concrete. In some areas these forms can be rented, often at a cost less than the cost of the lumber needed to build your own forms. This is worth checking into.

Green lumber works better than dry lumber for form work. This is because dry lumber absorbs large quantities of water from fresh concrete mix. Forms made of dry lumber, especially if they are large, have a tendency to swell, and if they are tightly fitted together (as they should be) they will buckle and warp. If kiln-dried lumber is used, the forms should be thoroughly soaked by spraying them with a hose before the concrete is poured. Better yet, after the forms are built, the inner surfaces should be liberally greased or oiled, to prevent the absorption of water and help them keep their shape, and to make the stripping easier after the concrete has cured. You can use linseed oil, automobile motor oil or something similar, or you can purchase a commercial form release oil. With small projects, wetting the wood with water is usually sufficient, and stripping is seldom a problem. The larger and more complex the project, the more helpful oiling is likely to be. The application of oil to patterned forms means a cleaner release when the forms are stripped and, consequently, a more sharply outlined imprint.

When you are building forms, keep three particular points always in mind. First, the forms must be extremely rugged and solidly set. Second, the forms must be virtually watertight, with no sharp cracks or gaps showing anywhere. The materials from which the forms are built must be free of knotholes and splits. And third, the inside of the form must faithfully imitate the dimensions and configuration of the project design. What the outside of the form looks like doesn't matter.

Concrete is heavy. The average weight is 145 pounds per cubic foot the range is from 130 to 160 pounds per cubic foot. In any

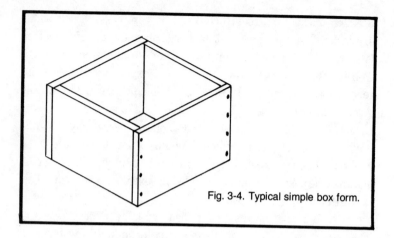

Fig. 3-4. Typical simple box form.

event, this means there will be a great deal of pressure exerted on the forms. Since the concrete mix is semiliquid, pressure is exerted not only on the bottom of the form, but also on the sides. The mix is always trying to escape, and has to be firmly contained. If there are holes in the form large enough for the mix to run out, you will run into problems. If a substantial amount of water leaks through cracks, the strength of the cured concrete can be impaired, at least in certain places; the finished surface might even be flawed by lumps and bumps.

Special care must be taken to provide plenty of support for the forms, especially if they are long or tall, so that there will be no bulging or sagging. If the forms do not remain perfectly straight, not only is the finished result unsightly, it may also be weak. These defects are caused by the development of a web of tiny cracks, especially in the case of sagging horizontal forms, on the surface of the concrete as it cures. This produces a poor finish and prevents the concrete from achieving its greatest potential strength. Moreover, the cracks will inevitably become larger, especially if exposed to any appreciable amount of weathering action, and the strength will be further diminished.

Small amounts of concrete, a few cubic feet or so, or relatively large concrete slabs which are not thick—say 4 inches or so—can be easily contained by boards of nominal 1-inch thickness (actually about 3/4-inch thickness). Solid bracing and heavy supports will keep the forms rigid and true. Larger masses of concrete demand proportionately heavier form construction. Nominal 2-inch-thick stock is most commonly used with two-by-six, two-by-eight, and two-by-ten planks to make up the form surfaces, with two-by-fours serving as supports and braces. Where the forms are large, it is often necessary to go to nominal 3- and 4-inch stock.

Box Forms

The most elementary kind of wood form is just a simple box shape as shown in Fig. 3-4. Sometimes these forms are made from redwood or cypress or some other treated wood and left permanently in place as a decorative part of the project. An example is a post base which will extend partly above grade. Plywood, in a 3/4-inch thickness (or even in a 1/2-inch thickness) for small jobs, can also be used. If the box's dimensions are not large, there is usually no need for additional bracing or support; the box is simply set into position, and the concrete poured. If the box shows any indication of bulging, pieces of two-by-two or two-by-four can be sharpened into stakes and driven into the ground at appropriate points as side supports (see Fig. 3-5).

Slab Forms

Slab concrete is easily formed up. Even though the volume of concrete to be poured may be considerable, the shallowness means

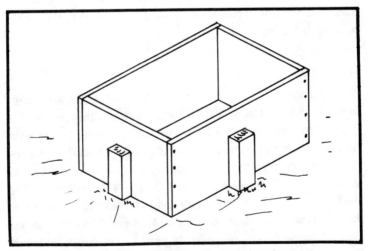

Fig. 3-5. Box form staked for extra strength and support.

Fig. 3-6. Typical staked and braced slab form.

that relatively little pressure is exerted at the sides of the slab. Two-by-fours, two-by-sixes, or whatever may be necessary up to a thickness of about 12 inches, may be set in place and securely staked, as shown in Fig. 3-6. More often than not, the thickness of the slab is dictated by the vertical dimension of the form rails. Thus, if two-by-fours are used as rails, the concrete is poured and finished level with the rail tops, resulting in a 3 1/2- or 3 5/8-inch slab thickness. Similarly, if two-by-sixes are used, the slab thickness will be about 5 1/2 inches.

Footing Forms

Footings, which are the support pads under walls, are usually handled in a somewhat different fashion. Just how they are handled depends to some degree upon the size of the footing. As a rule of thumb, you can figure that the width of a footing should be twice the thickness of the wall that it will support, and of the same thickness as the wall (Fig. 3-7). In other words, if the wall is 8 inches thick, then the footing should be 8 inches thick and 16 inches wide. If the wall is 12 inches thick, then the footing will be 12 inches thick and 24 inches wide. This rule is the one generally used for residences and small buildings where the above-footing mass is not particularly great.

In the case of garden walls, patio edge footings, and similar applications where the loading on the footing is negligible, the footings can be made smaller. Here you can figure the width of the footing as equal to the thickness of the wall it will support, or from 2 to 4 inches wider. For walls of 3 feet or less, or for perimeter footings of large slabs, a thickness of 6 inches is usually sufficient. For walls up to about 5 feet in height, a width of half again the thickness of the wall plus an additional 2 inches of thickness should be enough. For calculating footing requirements of walls exceeding 5 feet, it's just as well to revert to the use of the previously mentioned rule of thumb. Though the loading is still minor, you are allowing for a substantial base to the wall, a factor which is especially helpful in the case of freestanding walls. Some footings can be formed and poured at the same time as the structures they support (Fig. 3-8).

To make a wood footing form, dig a trench that is comfortably wider than the footing; this is to give yourself plenty of room in which to work. Position nominal 2-inch-thick planks on edge for the sides of the form. Line the planks up carefully and level them; be sure to check both alignment and level continuously throughout the con-

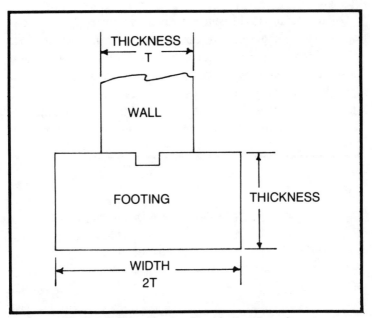

Fig. 3-7. Relationship of wall and footing dimensions.

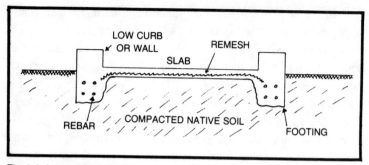

Fig. 3-8. Integral footing, slab and curb structure poured directly on compacted soil in one operation.

struction of the form. Drive a series of two-by-two stakes against the rails at appropriate locations, as shown in Fig. 3-9. Don't nail the rails to the stakes, just prop them into position. Nail a series of spreaders across the rail tops to hold them in place. If the bottom edges of the rails tend to cant inward, you can hold them fast against the support stakes by wedging small sharp rocks along the inside bottom edges, or by driving spikes at an angle into the ground. If necessary, you can add secondary stakes and wedges for extra strength.

If the footing is to be poured in a narrow trench where you don't have much room to work, you can use one of the methods shown in Fig. 3-10 to set the forms. With both methods, it is generally easiest to set the rails in place, roughly aligned and leveled, and then nail spreaders at fairly close intervals along the top edges. Place more spreaders at intervals along the trench bed to insure that the rails don't buckle inward at their bottom edges. Don't nail the spreaders;

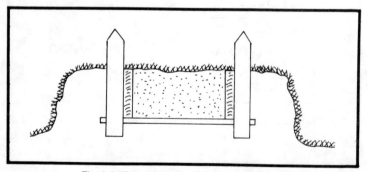

Fig. 3-9. Typical simple staked footing form.

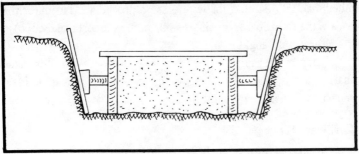

Fig. 3-10. Footing form propped and wedged in trench.

they're only serving as temporary props to maintain the right dimension while you drive the stakes, or set the side wedges in place. As the form approaches the completion of the leveling-and-alignment stage, remove the bottom spreaders. If the sides of the forms still want to kick inward, restrain them by driving spikes into the ground alongside, or by wedging short lengths of iron rod or reinforcing bar between the rails until their tips extend only an inch or two above the trench bed. These can be left in place when the concrete is poured.

If the footing must make a right-angle corner, follow the general layout pattern shown in Fig. 3-11. If the corner angle is something

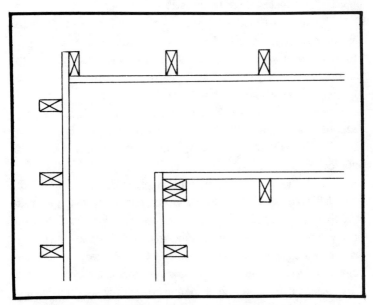

Fig. 3-11. One method of making a sturdy right-angle footing form.

other than 90 degrees, follow the same procedure, but miter the ends of the form boards (perhaps some of the support stakes, too) at the intersecting angle.

Oiling or greasing the rails is not usually necessary for these shallow forms, though soaking them with water is a good idea, especially in hot, dry weather.

Curvilinear Forms

Curvilinear forms, such as might be used for a meandering walkway or free-form garden edging, or at times even for footings, are put together in the same pattern. There *are* differences, though: the material used for form rails is different; additional support must be used; and extra care taken to ensure a good job. Any straight sections of appreciable length can be formed in the usual manner for slabs or footings. The curved sections can be made from any of several different materials, but must be flexible enough to be bent into the needed radii. For very small jobs where the slab is thin, 1/8-inch Masonite cut into long strips works fine. Coil or roll sheet metal, either steel or aluminum, will also do the job, and so will corrugated aluminum garden edging. The latter, of course, will impart a corrugated edge to the slab.

The same materials can be used for slabs up to about 6 inches in thickness, except that the Masonite should be increased to 1/4-inch stock. Exterior grade plywood also makes good form rails. As the thickness of the slab or footing rises to a foot, the rails must be sturdier. Heavy-gauge sheet metal can be used, but is expensive and somewhat difficult to work. Two layers of 1/4-inch Masonite or exterior-grade plywood will do the job, with the two layers being placed into position at the same time, but not attached to one another.

Forms should abut one another snugly at the joints. When the joints fall along the radius of a curve, care must be taken to make sure that the arc remains true in curvature, and does not flatten out or develop a salient. A great number of support stakes have to be driven to hold the form rigid and keep the rails from waving and bulging. Special attention has to be given to circular forms, so that the finished edge of the slab or footing does not weave and wander. Small stakes made from nominal 1-inch stock, about 1 1/2 inches wide, are usually ample and are easy to make. Or, you can buy ready-made stakes (called grade stakes because they are used by surveyors to set grade lines and elevations) at most lumberyards. Incidentally, when you're finished with these stakes as form supports, you can clean them off, turn them pointed end up, and make a low picket-fence garden edging. Just nail a whole series of these stakes to a double rail made from 16-foot lengths of 1/4-inch lattice stock. The lattice rails are flexible enough to follow the curved lines of a free-form garden, too.

For heavy slabs or for footings, using larger stakes is a good idea. You may find it necessary to double-stake, and also to put in extra bracing (run at about a 30-degree angle from the bottom of the rear stake to the top of the stake directly against the form rail). In the case of footings, stretchers should also be installed across the top of the form, running from stake to stake. Again, the support stakes should be closely spaced along the length of the rails, and the line of the arc carefully set. In all cases where Masonite or plywood is used for form rails, oiling or greasing is a good idea to prevent water absorption. Sheet metal rails need no treatment.

Wall Forms

Poured concrete walls of various types are often used around homes, either for privacy or for decorative purposes. These walls may be left plain, or be imprinted with some decorative design, or painted, or veneered with some other material. The forms for such walls are considerably more rugged than those previously described, are more complicated, and have to be carefully constructed. The walls of the form are generally made from nominal 1-inch planking, and sometimes of 2-inch material. Stakes are not used, but vertical studs resting either upon the ground or upon base shoes located along the bottom of the wall serve the same function. To hold the studs in place, inclined bracing, run at a 30-degree angle, is used in concert with horizontal bracing at ground level. Both types of bracing must be tied to sturdy stakes. The matter of bracing is variable, and depends upon the circumstances. The object, however, is to make the form sides as sturdy as possible, and this can be done by whatever means best suit the situation. The studding is further anchored and reinforced by horizontal braces called *wales*. A

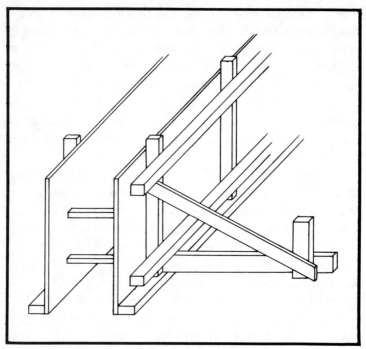

Fig. 3-12. Typical braced wall form.

series of spreaders is placed within the form, and the form is snugged tighter by the use of ties, or combination tie/spreaders.

There are many alternative methods of making wall forms. Professional concrete workers may use 3/4-inch plywood, or special heavy, treated form board, or steel forms which lock together. One of the latest innovations is a form made from foam plastic which is locked into place in panels and sections, filled with concrete, and then left permanently to serve as insulation for the completed structure. But in most home projects, unless you're building the house itself, it is not necessary to go to these lengths. A typical wall form is shown in Fig. 3-12.

Wall forms must be both rugged and accurately assembled, although, in general, the method of construction can be reduced to doing whatever happens to be easiest and most satisfactory for the job. After all, building a wall form isn't the same as building a cabinet. If the interior surfaces of the form are true, clean, and watertight, that is all that really matters.

The spreaders are installed at various heights and at appropriate linear intervals to insure that the walls of the form remain the proper distance apart. These spreaders needn't be attached solidly in place, since they must be removed at some point during the pouring operation. If it's a low wall, you have only to reach into the form and lift out the spreaders as the pouring proceeds. However, if the wall is a high one, then you will have to work out some method of retrieving the spreaders. One method that works well is to reach into the forms with long-handled fireplace tongs. Another possibility is to drill holes in each but the lowest unit in a vertical series of spreaders for passing down a pull-wire for attachment to the bottom member, as shown in Fig. 3-13. As the concrete is poured, the spreaders are successively yanked up out of the way until they are all clear of the form.

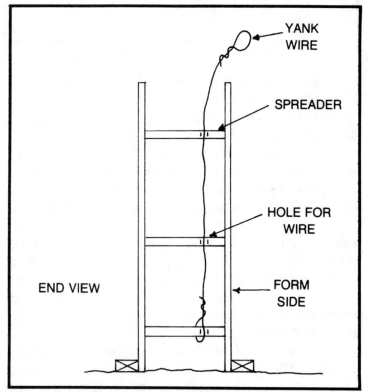

YANK WIRE

SPREADER

HOLE FOR WIRE

END VIEW

FORM SIDE

Fig. 3-13. Arrangement for easy removal of spreaders from wall form by means of yank-wire.

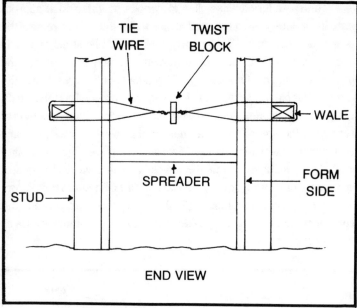

Fig. 3-14. Tie-wire installation in wall form.

The forms are tied together as an additional safeguard against bulging when they are subjected to the weight of the concrete. As the form is built, the ties are drawn up snug (but not too tight) against the spreaders to maintain proper dimensions. When the concrete is poured, the mix itself takes the place of the spreaders and exerts pressure against the ties, but is restrained from pushing outward. The ties remain in place during the pour, and are not removed. One easy method for setting ties is shown in Fig. 3-14. Soft black iron wire, which is often called utility wire or mechanic's wire and is available at hardware stores, lumberyards, and automotive supply houses, is about the best for this purpose. A thickness of 8- or 9-gauge is ideal for most applications. If this type of wire is not available, you can substitute the ordinary barbed wire which is used for livestock fencing, or plain steel stock-wire. Both types, however, are difficult to work with.

To make a tie, drill small holes in the form sides and loop a length of wire through the holes and around the nearest wale or stud, and join the loop ends. Insert a 20-penny nail or a spike in the center of the wire and twist until the form sides are drawn snug against the ends of the spreader. When the time comes to strip the forms, clip the wire on each side, remove the forms, and snip off the protruding pieces of wire close to the surface of the cured concrete. Now, the exposed wire endings will cause rust streaks on the concrete surface after a time. If this is objectionable, chip out a small hole around the wire and trim the wire off below the surface of the concrete. Then patch over the hole with a fine grout to seal off the wire completely from the weather.

Several types of commercial combination tie/spreaders are available. Though the styles differ, the general idea is to run a specially-formed bolt-like affair through the form. The rigidity of the rod holds the form walls to the proper dimensions; but because the rod is anchored to the studs or wales, or both, it remains solidly in place as the concrete is poured and does not allow the form to bulge outward. When the forms are removed, the rods are unbolted and broken off (at a manufactured stress-point along the shaft) just below the surface of the concrete. The blemishes can then be grouted over if desired.

One of the problems involved in making large wall forms is determining the proper spacing of studs and wales along the form side, so there is no chance of the form bulging or springing apart. There are three factors involved here. The weight of the concrete mix itself has already been mentioned, and this weight can be considerable when a 6- or 8-foot high wall is involved. But the rate at which the mix is poured into place also plays a part, and so does the temperature of the concrete. These factors in combination are responsible for the *hydrostatic pressure* on the forms. The lower the temperature of the concrete, and the faster the rate of placement, the greater the hydrostatic pressure that will be developed within the form. This pressure, incidentally, is greatest at the bottom of the form and least at the top. If you pour a 6-foot-high wall in an hour's time, using a concrete mix with a temperature of 40°F, the pressure developed may be as high as 1,500 pounds per square foot (Fig. 3-15). This means one thing in particular—that form has to be mighty solid.

For a given amount of pressure exerted, the distance between studs is dependent upon the thickness of the form walls, and vice versa. With that same pressure of 1,500 pounds per square foot, for

Concrete Placement Rate, Feet Per Hour	Approx. Concrete Pressure in Lb. per Square Foot at Concrete Temperature of:			
	40°	50°	60°	70°
1	490	440	390	350
2	700	600	520	460
3	900	760	660	590
4	1100	920	800	700
5	1300	1080	930	820
6	1510	1240	1060	930
7	1700	1400	1200	1050
8	1790	1450	1250	1190

Fig. 3-15. Relationship of concrete temperature, pouring rate and pressure.

instance, form walls made of 1-inch stock would require a stud spacing of every 12 inches. If the wall thickness were increased to 1 1/2 inches, then the stud spacing could safely be increased to about 20 inches (Fig. 3-16). Similarly, the spacing of the wales is depen-

Concrete Pressure, Lbs. per Square Foot	Approx. Stud Spacing in Inches With Sheathing Thickness Of:			
	1″	1 1/4″	1 1/2″	2″
200	24	32	38	--
400	19	26	31	38
600	16	22	27	34
800	15	20	25	31
1000	14	19	23	28
1200	13	18	22	27
1400	12	17	21	26
1600	12	16	20	24
1800	11	15	19	23

Fig. 3-16. Table of form support stud spacing.

dent upon the thickness of the form walls and the size of the studding, since the wales serve to reinforce both. Thus, if two-by-four studs are used in conjunction with 1-inch sheathing, and the pressure is 1500 pounds per square foot, then the maximum wale spacing should be about 10 inches (Fig. 3-17). The heavier the studding and the thicker the form walls, the farther apart the wales can be spaced.

Column Forms

One of the most widely used configurations of concrete is the column. There are several approaches to building the forms for concrete columns. The simplest and most effective approach, whether the column is round or in the shape of a rectangular pier, or in the case of footings for columns made either by a separate concrete pour or from another material such as steel or wood, is to pour the concrete directly into an appropriately shaped excavation in the ground. The earth itself acts as the form.

Where the column has no separate footing and yet rises above grade level, then some forming must be done. One option is digging a hole for the below-ground portion of the column, and then building a wooden form for the above-ground portion.

Concrete Pressure, Lbs. per Square Foot	Approx. Wale Spacing In Inches With Stud/Sheathing Combinations of:				
200	48	33	--	--	--
400	32	24	--	54	45
600	24	18	57	44	36
800	20	14	48	36	30
1000	17	12	42	31	26
1200	15	11	38	27	23
1400	14	10	33	24	20
1600	12	9	30	22	19
1800	--	--	27	20	17

Fig. 3-17. Table of form wale spacing.

Fig. 3-18. Pier form of heavy cardboard tubing filled with concrete.

Another method is to dig a relatively deep hole wide enough to accept a commercial cardboard form tube. This tube consists of spirally-wrapped layers of heavy cardboard and is available in a range of diameters and lengths. After purchasing a length of tube, trim it off as needed and set it in the hole. Brace it in place carefully with rocks or chunks of scrap wood. Pour the concrete directly into the tube (Fig. 3-18). After the concrete has cured, backfill and tamp the hole and peel away the cardboard from the above-grade portion of the column.

Another method is to use sections of clay flue tile. The first section or two, depending upon how far underground the column will extend, are set in a hole; the remaining sections are then stacked on top and their joints mortared. When the mortar has cured, the tile is poured full of concrete. Vertical pieces of reinforcing bar can be stood upright within the tile section for added strength. Concrete blocks with large cores also make effective piers, and can be set directly in the earth, below grade at an appropriate depth, or upon a below-grade footing. The blocks are mortared together in the usual fashion, and after the joints have cured, the hollow cores are filled with concrete. Here again, reinforcing bars, for added strength, can be set into the cores. Actually, practically anything that is hollow, strong enough, watertight, presents a suitable appearance or is removable after the concrete cures, can be used for column or pier forms.

Wood forms for columns, column footings, piers, pier footings, and combinations where the footing and the pier or column is poured at the same time can be easily made much in the manner of wall footings. Figure 3-19 shows a typical form of this sort. External bracing is usually not necessary; the form is self-contained and self-supporting (although it must be rugged enough to withstand the concrete pressure). Props, braces or wedges, or combinations of all three, may be necessary to level and align the form and hold it securely in place so it cannot drift around during the pouring operations.

Anchors

In many small projects, posts can be cast directly into the column or pier, as in setting a corner post, fencepost, flagpoles, or a bird feeder pole. As the concrete is poured, the post is placed in position and securely guyed; the mix is poured around it. In many other cases, though, as with posts and columns and with plates applied to a wall top, or where some sort of superstructure must be

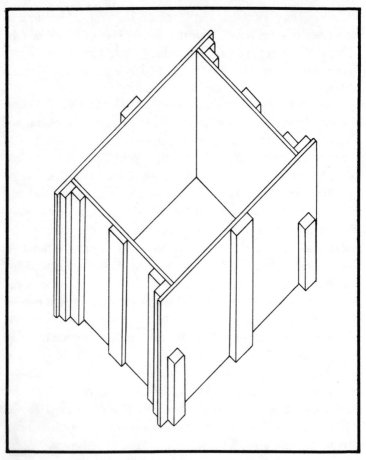

Fig. 3-19. Typical pier or column footing form.

attached to the concrete, anchoring devices must be installed in the concrete. This is sometimes done by positioning the anchors within the forms, and then pouring the concrete mix around them. In other instances, it is possible to insert anchors into the freshly poured concrete during the finishing process. Much depends upon the method of anchoring, and there are many.

One method consists of using long, heavy steel bolts with large steel washers or square, drilled plates of steel resting next to the bolt heads. The bolts are positioned upside down in the concrete mix, with an appropriate length of thread protruding above the finished surface. Hook-type anchorbolts, which are made especially for this purpose and come in various sizes and diameters, are set in much the same way. There are also various kinds of saddle or bracket devices used to anchor wood or steel columns or posts. Some of these devices are partially embedded in the concrete, while others are fastened to the cured concrete surface with bolts which have been accurately emplaced in the concrete mix. Chunks of ordinary angle iron can be attached with bolts preset in the concrete, with the square or rectangular posts later bolted to the vertical portions of the angle.

One of the simplest methods of securing wooden posts which will have a sufficient top load on them to hold them solidly in place is to use a center pin. A long bolt is embedded in the concrete at the proper location, and a matching hole is drilled into the end of the post. The post is simply dropped over the pin as construction proceeds. If it is important that the post not be able to turn at all, use a pair of pins. Various types, brands and designs of anchors are available at lumberyards. The best policy is to find out what kind of anchors can be purchased with a minimum of fuss and bother, and then adapt your anchoring designs to that particular hardware.

Step Forms

One type of form which requires a little extra care in the construction is that used for pouring concrete steps. Most steps found around residences are the earth-supported type, by far the easiest type to build. Though the forms are relatively simple, they must be accurately dimensioned, ruggedly supported, and carefully put together if a top-notch job is to result. Fig. 3-20 shows the forms for a set of steps such as might be employed for connecting different levels in a garden, for instance, or as an approach to an elevated patio or terrace. Here the entire set of steps is poured as a single large block of concrete. The sides could be left as is, veneered with stucco or stone or even tile, or the whole affair could be set back into the upper level (instead of protruding) and completely surrounded by earth or other masonry. Figure 3-21 shows a somewhat different arrangement where the side retaining walls are poured first, and the steps themselves afterwards (with tamped earth as a base).

The riser and tread dimension combinations used for outdoor steps are generally different than those for indoor steps: broad

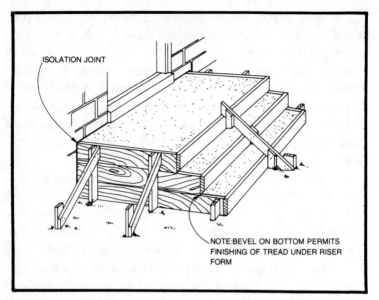

Fig. 3-20. Typical form for steps. Courtesy of the Portland Cement Assoc.

treads and low risers are preferable outdoors. An excellent riser height is 7 1/2 inches, which means that you can use nominal 8-inch stock lumber in the form. A proportionate tread width for a 7 1/2-inch riser would be 10 inches. The minimum size for a tread should be about 9 inches and the maximum about 12 inches. If broader treads are desired, make them considerably broader, and not just a little. They should be wide enough so that a user is aware that he must make a two-stride advance on each level. There is an awkward middle distance involved here, somewhere between 12 inches and 16–18 inches, where the step is really too wide for one pace but too narrow for two, and the unwary user can stumble all too easily.

The simplest type of tread/riser construction leaves the outside intersection perfectly square. However, only a little extra work is involved to mold the riser for a more attractive appearance. Figure 3-22 shows some of the ways that this can be done, and your imagination can doubtless conjure up others. If railings or balustrades are to be installed later on, don't forget to set anchors for them. You might also want to extend the bottom tread to one or both sides, and perhaps the top tread as well, to provide support pads for

Fig. 3-21. Steps formed between existing side walls.

the railing. This would probably be essential if the railings are to curve or curl outward at each end, as many ornamental types do.

Also, note that the treads should not be perfectly flat. There should be a slight forward pitch of about 1/16 inch so that water will

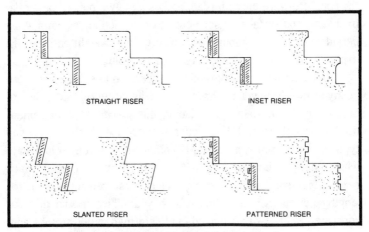

STRAIGHT RISER

INSET RISER

SLANTED RISER

PATTERNED RISER

Fig. 3-22. Various methods of forming step risers.

drain off. Take care that the pitch does not exceed this amount, however, as too much of a pitch can be dangerous for the users. A backward pitch is also undesirable, because water and ice would collect at the tread/riser joint. You can install this pitch during the finishing process just by eye, by building the concrete up a tiny bit above the form at the rear of the tread, or by sloping the form sides as you build them. The treads can be skidproofed during the finishing process by roughening or patterning the surfaces, or by embedding pebbles or grit in the uncured concrete surface after finishing.

REINFORCING

When you have completed the form, you'll have to turn your attention to inside-structure reinforcement. Concrete is pretty solid stuff, but its strengths are far from being equal in all aspects. Concrete's compressive strength, for instance, is tremendous. An incredible amount of force is required to crush a chunk of concrete that is fully supported. Poured concrete is extremely durable and resistant to all kinds of hard blows. But the tensile strength is comparatively slight. If you were to cast a slab of concrete a couple of feet wide, 4 feet long and 4 inches thick, prop up each end with a piece of two-by-four, and then jump hard onto the middle of the slab, chances are that you would break it cleanly in two. About the same thing could happen if you were to make a driveway of the type that uses two separate concrete runways just wide enough to drive on. If the concrete were unreinforced, and especially if poured on a relatively soft soil surface or subjected to frost action, within a short period of time the combined action of vehicle weight and thermal expansion/contraction would break up the slabs.

And there are further problems as well. Concrete does not cure nearly as rapidly as you might think. The initial curing rate is dependent upon several factors, including the specific type of cement used, the weather, and the amount of water used in the mix. Some types of concrete reach their full compressive strength in 1 day, while others require as long as 3 months. But even types which reach a high compressive strength quickly can remain "green" for lengthy periods, often as much as a year. This means that the concrete can be easily damaged (and deformed under heavy loads) even though it appears to be perfectly solid. Concrete, because it

OLD	NEW	
1/4″	#2	
3/8″	#3	
1/2″	#4	
5/8″	#5	Fig. 3-23. Reinforcing bar designations.
3/4″	#6	
7/8″	#7	
1″	#8	
1″ sq.	#9	
1 1/8 sq.	#10	
1 1/4 sq.	#11	

retains its plasticity for months—depending upon the conditions—will actually creep a small amount during the curing period.

The answer to these problems, at least in part, lies in the use of reinforcing materials. There are two materials in particular which are of interest to the do-it-yourselfer. One is reinforcing bar, generally shortened to *re-bar*; the other is reinforcing mesh, usually called *re-mesh*. The latter is also called welded wire screen.

Reinforcing rod which was perfectly round used to be popular, but in the interest of better bonding, the type now favored is *deformed bar*. This is available in a number of sizes which were once designated by diameter in fractions of an inch, but now go by a scale of numbers (Fig. 3-23). Re-bar comes in standard lengths, but most lumberyards will sell it by the foot, and also cut it for you into the lengths you need.

Re-mesh is bought by the roll in standard widths and lengths, and then cut with a hacksaw or heavy-duty wire cutter to fit the specific application. A few suppliers sell re-mesh by the running foot.

The proper use of reinforcing materials in poured concrete requires a great deal of complex engineering, especially in large projects where complicated stresses and strains are involved. Fortunately, the do-it-yourselfer doesn't have to worry much about all of this. If you make it a rule to include reinforcing of one sort or another in all of your concrete work, then you will be safe. The finished product will be tougher and stronger and better able to withstand wear and tear and the passage of time than would otherwise be the case. The added amount of work is minor, and the expense negligible. All in all, adding reinforcing to your concrete projects is well worth the trouble.

Re-mesh is probably the more widely used reinforcing material for small projects of the home variety, because it is easier to work with, less expensive, and in most applications adds more than adequate strength. Re-bar is more commonly used in applications where re-mesh would be difficult or impossible to install, such as in thin columns, or in narrow piers and footings. The placement within the concrete of either type is largely a matter of common sense, but there are a few pointers to follow.

Re-mesh is used in all slab pours, such as driveways, walks, patios, pools and the like. Where the thickness of the section is about 4 inches, keep the mesh about 1 inch above the ground. If the section is thicker, say about 6 inches, bring the mesh up to about 2 inches off the ground. The actual installation is a matter of unrolling the sections of mesh and cutting them to fit. Wherever more than one run is necessary, or where small pieces of mesh must be joined, make sure that they overlap 6 inches or so, and lock the pieces of mesh together either with wraps of mechanic's wire or by entwining the cut ends of the mesh itself. Be careful when handling the mesh as you take it from the roll; it can be mean. It has a tendency to recoil into its roll form, and can snap back and puncture your hide.

Stretch out the pieces of mesh on the ground and walk around on them until they are as tractable and flat as you can reasonably get them. After they are set in place, prop them above the form bed with small washed stones to achieve the proper ground clearance. Another method, handy when the project is so large that you have to walk on top of the mesh during the pouring process, is to leave the mesh flat on the ground, and as the concrete is poured, reach down into the mix with a hook and raise the mesh an inch or so (the larger aggregate will hold it there). You really don't have to worry much about being dead accurate as far as the mesh's distance above the ground is concerned, and the mesh will remain embedded in the concrete mix at exactly the level you position it with the hook.

Mesh also works nicely in wall sections. Flatten pieces of mesh of the right size, and hang them into the forms like curtains, just about in the longitudinal center of the form. This can be done by suspending the mesh from wood hanger-bars set across the form top, and tying loops of mechanic's wire to both the bars and the mesh. As you begin the pour, keep an eye on the bottom of the mesh curtain to make sure that the concrete does not move it out of line. As the concrete rises in the form, the mesh curtain will be fixed more and more rigidly into place. When the pour is complete, clip away the mechanic's wire and remove the hanger bars.

In all cases when using mesh, leave a sufficient margin around each edge so the mesh does not protrude from the finished concrete. No matter how persevering you are, you can never get mesh to lie absolutely flat, and there will always be some waviness and irregularity in its spread. This, however, is not a problem, even though the mesh-to-ground clearance will be somewhat variable, and the wire will not be evenly embedded. It does no harm, and there won't be any loss of strength.

The small quantity of re-bar that is likely to be used in most home projects can be installed either before or during the pour. Choose whichever method is more convenient; in either case, the re-bar will bed into the concrete perfectly well. As far as size is concerned, numbers 3, 4, and 5 are probably the most often used in small home projects. Number 3 is more than adequate, while number 5 can be just as easily used, but is a bit more expensive and probably will impart greater strength than actually is necessary. Also, understand that you can achieve the same effect by using a large quantity of small re-bar or a small quantity of larger re-bar.

The number 3 size is easy to handle, and is very effective when installed in a grid arrangement. For projects which will carry a relatively heavy load—such as the base for a barbecue—a grid of number 3 bars spaced 6 inches apart in both directions is about right. For projects which will bear lighter loads, the mesh spacing can be increased to 12 inches. To make a grid of this sort, the usual procedure is to lay out the bars in crisscross fashion and cut to the right lengths, then tie all the pieces together with mechanic's wire where they cross. The grid is then placed at about the mid-level of the pour. In other words, for a 4-inch slab you would pour the first 2 inches, then set the re-bar grid in place and jiggle it a bit to bed it firmly into the fresh concrete. Then finish the pour.

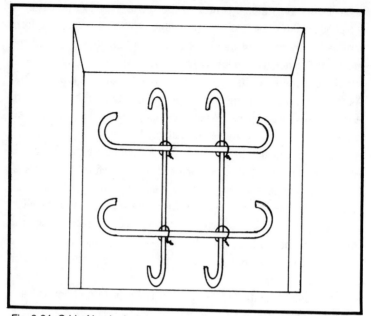

Fig. 3-24. Grid of hooked rebar set in excavation for pier or column footing.

Re-bar installations for other types of work are accomplished in much the same way, by making a grid of appropriate size and then setting it in place. Where the projects are large, you may have to install the re-bar a piece at a time, wiring the intersections together as you go. If the installation is made before the concrete is poured, the re-bar must be supported at the proper point above the ground either by propping from below with stones, or suspending from above with wires. As with the mesh, care must be taken to ensure that the re-bar fabrication does not get knocked out of line during the early stages of the pour.

Re-bar can also be easily bent to follow curves and angles, or shaped into hooks for extra strength. The arrangement shown in Fig. 3-24 is typical of a re-bar grid used to reinforce a footing for a column or pier. If there are no forms from which the grid can be suspended, you can prop up the grid with washed stones, pieces of concrete blocks or bricks; alternatively, you can devise your own suspension system. Park a couple of saw horses near the form, but not in the way of workers, and run a couple of two-by-fours across the horses. Then suspend the grid from the two-by-fours with wire loops. Just make sure that the grid stays where it is supposed to as you pour.

Either re-bar or re-mesh can be installed in columns and piers, too. Exactly how this is done depends largely upon the size and shape of the structure. Re-mesh can often be curled into a tight cylinder and stood in place; the concrete is then poured around the cylinder. For square structures—perhaps a pier—some judicious bending of the mesh will produce a roughly square-sided tube which can be either stood or suspended in position. Bars can also be wired together with heavy steel or iron wire to form tubular arrangements of circular or square cross section for setting into the form. Sometimes, where quarters are close, you may have to slide individual reinforcing bars into position as the pour is made. Such would be the case, for instance, in a pier made form concrete blocks; afterwards the cores would be filled with concrete or grout. One or two rods can be slid down into each core after the pouring has begun. They are stood in place, and then concrete is poured around them to complete the job.

There are a few applications, as far as household projects are concerned, where reinforcing is not necessary. A case in point would be the instance in which concrete is poured in a hole in the ground to serve as an anchor for a fencepost. Not only is there very little room for reinforcing, but the added strength is not necessary anyway. Wall sections are generally left unreinforced, too. In most other cases, however, reinforcing is advantageous.

Here are some generalities to follow as you consider the use of reinforcement for your concrete projects.

First, the reinforcing is usually installed only in the areas where tension stress, represented by a load, is the greatest. As a rule, these points will occur close to the surface directly opposite the load. Where a load is more or less evenly distributed upon the concrete section, as in the case of a footing, for instance, the reinforcing material is also evenly arranged. In instances where the load won't be concentrated at any specific point, as in the case of a walkway or patio, the slab can be reinforced over the full area—to add strength and reduce the possibilities of cracking, shifting and buckling or settling from frost or other action.

Second, the reinforcing material should always be surrounded by a liberal quantity of concrete. A general rule of thumb is to make sure that no less than 3 inches of concrete surrounds the reinforcing materials on all sides when the mix is poured without benefit of forms, as in an earth trench. Where forms are used, no less than 2 inches of concrete should surround all reinforcing members. The exception to this is the use of mesh in thin slabs, where the mesh can be embedded as little as 1 inch above the ground. In the case of concrete structures which won't be poured on the ground or exposed to weather of any sort, the minimum thickness of surrounding concrete can be reduced to as little as three-quarters of an inch.

Third, if you bend reinforcing bar into hooks, curves and other shapes, do so without the use of heat. All re-bar should be bent cold.

Fourth, make sure that all reinforcing bars are clean and free from rust, mud, paint, oil and grease. Only clean iron will properly bond to the cement. Admittedly, this is not a hypercritical issue with most home projects, as it can be in commercial concrete work, but nonetheless the point is worth bearing in mind. Neither is it particularly critical with respect to reinforcing mesh, which is frequently rusty. But even so, it is wise not to use excessively rusted material. Dirty or mud-caked mesh can be cleaned with the garden hose.

And last, the easiest course to follow when buying reinforcing bar is to work out all the dimensional details ahead of time, and present your supplier with a list of the exact pieces you need. Then let him do the cutting. Most suppliers have special cutters for this purpose, and even if there is an extra charge, this is a whole lot easier than endlessly trying to gnaw off pieces of re-bar with a hacksaw.

KEYS

Keys are used when a project is to be accomplished by way of several separate concrete pours, and each pour is allowed to cure before the next pour is made. Suppose, for instance, that you want to build a concrete wall around your building lot. By doing the job in several sections, you can work out a manageable schedule and not be faced by a horrendous amount of work all at once. You will need a far smaller amount of form material, the overall cost will be lower, and the finished job will be just as good as if you made a monolithic pour. But, each section must be keyed to the next.

This is a simple matter, and is taken care of during the form-building stage. When you make the end wall of the form section, nail a strip of wood along the inside vertical centerline. For a wall up to about 8 inches thick, a length of two-by-two works fine. For thicker walls, use a piece of two-by-four, laid flat. Figure 3-25 shows how this is done. When the forms are stripped, you'll have a *keyway* molded into the end of the section. Then when the next pour is made, the two sections will be automatically keyed. This greatly increases the strength of the walls, reduces the possibility of cracking, and curbs the tendency of wall sections to tilt out of line with one another.

Another instance in which a key is widely used is when a poured wall is joined to a footing, as shown in Fig. 3-26. In this case, the footing is poured in the usual fashion, and during the finishing process a strip of wood (usually a two-by-four) is pressed down into the soft concrete. Remove the strip before the concrete sets up. The top edges of the keyway may crumble a little bit, but this is of no consequence. When the wall is subsequently poured atop the footing, the key is automatically formed and the footing and wall for all practical purposes become one immovable unit, well bonded and with little chance for shifting.

By employing this general principle, you can provide keys and keyways of appropriate dimensions for just about any project that is to be made with more than one pour.

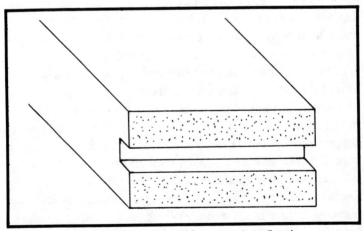

Fig. 3-25. Keyway formed into concrete wall end.

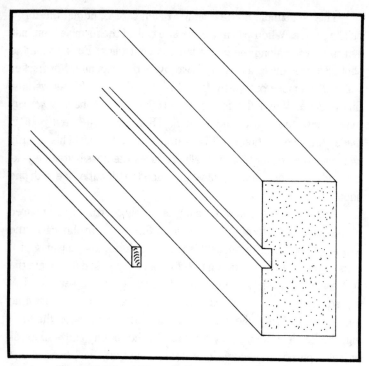

Fig. 3-26. Keyway formed into concrete footing top.

TIES

Ties perform a function similar to that of keys and keyways, and are even more sturdy and effective, though additional labor and expense is usually required. Though most ties are from concrete to concrete, they may also be from concrete to some other object or structural element which must be rigidly embedded.

Assume, for example, that you want to plant a length of 1-inch pipe into a concrete anchor or pedestal to serve as a mount pole for a bird feeder. The pipe is smooth and straight-walled, and there isn't much surface area to which the concrete can bond, even if you set the pipe a couple of feet deep. Furthermore, there is the problem of unequal expansion/contraction between the two different materials which is bound to take place. All of this means that the pipe can easily loosen up. However, if you drill a hole or two through the pipe at the anchored end and insert a couple of long nails or bolts through the holes (Fig. 3-27), you will have provided ties that the poured concrete can grip solidly. The pipe will never come loose. Similarly, if you drive a few 20-penny nails into the bottom end of a fencepost before pouring concrete around it, that post will never move.

Concrete-to-concrete ties are handled in much the same fashion. The object is to lock together two adjoining sections of concrete, one of which was poured and cured before the other. The purpose is the same as with a key, but the result is even more effective. One possibility is to use curved or hooked anchor ties as shown in Fig. 3-28. Here the bottom half of each hook is embedded in the footing during the finishing process. When the wall is later poured atop the footing, the two are solidly locked together. Ordinary heavy bolts with washers or drilled plates at each end will accomplish approximately the same purpose, and so will emplanted pieces of reinforcing bar, which may be bent or not as you desire. Straight pins cut from re-bar can be set at angles which oppose each other (rather than being positioned perfectly straight up and down).

Another possibility arises where reinforcing bars are used extensively in the first pour. Instead of completely embedding all of the bars, certain pieces are left extra long and are bent so that they stick up from the finished slab or section. Here again, the exposed ends may be bent into hooks or angles for added strength. The subsequent pour covers all of the protruding rods, and the two units are locked into one.

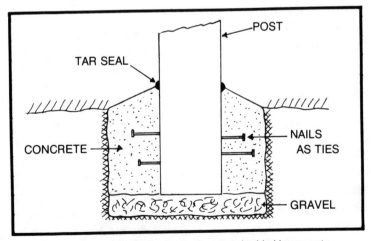

Fig. 3-27. Nails set as ties in wood post embedded in concrete.

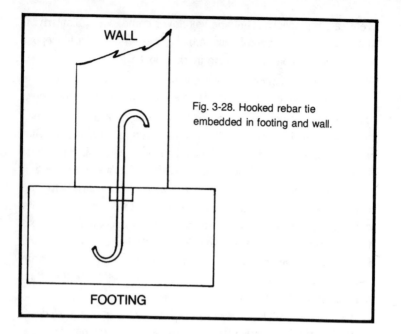

WALL

Fig. 3-28. Hooked rebar tie embedded in footing and wall.

FOOTING

This same system can also be used to join a concrete footing to a masonry unit wall built from brick or concrete block or even stone. Here the protruding reinforcing bars are arranged so that they will fall within the brick or block cores, or in the cracks between stones. They are mortared firmly into place as the wall construction proceeds.

GROOVES AND JOINTS

Grooving is a process that is used most often on walkways, but can also be employed on other slab-type applications as well. This is done with a special tool which incises a round-edged and sharp-bottomed groove about an inch deep. Transverse grooves are created along the walk at regular intervals. Though there is a decorative aspect to grooving, the principal reason for doing it is to avoid unsightly and sometimes dangerous random cracking which could occur if the grooves were not present. Concrete walkways, as well as other slab constructions, merely float upon the earth. In cold climates, freezing and thawing beneath the slab takes place constantly, and the slab itself expands and contracts in response to the wild temperature gyrations. In practically any locale, a certain amount of heaving and settling is virtually inevitable because of tree roots, excessive moisture runoff, or for any of a hundred reasons. If the slab is properly grooved, whatever cracking does occur is likely to take place along the line of the groove, since this is the weakest point in the slab.

Grooving can be used for decorative purposes on larger slabs. The patterns can be made randomly or symmetrically in whatever fashion the builder chooses. A grooving tool can be used for this, but there are other implements that would cut the desired kind of groove just as well. Again, if cracking does take place, most of the action will occur at the grooving lines rather than in the open expanses.

Expansion joints should be placed between new concrete slab construction and abutting structures (previously poured concrete slabs, existing curbs, lamp poles, fire hydrants, masonry walls, steps, building walls, etc.). You can purchase expansion joint material or you can cut strips from 1/2-inch-thick fiberboard. The expansion joint material should be set in place and secured before the concrete is poured.

Expansion joints may also serve a decorative function by breaking up a large slab area into a number of smaller ones. In that case, the expansion aspect isn't all that advantageous but it causes no harm, and, after all, the decorative aspect is the main point. A grid system made of redwood lumber is frequently used for this purpose, either in a nominal 1-inch or 2-inch thickness. The lumber rests edgewise on the slab bed. This makes the pouring easier, since the job can be done in a series of small steps rather than one great big one. Other types of wood may be used also, provided they are properly treated to withstand the ravages of weather and insects.

PREPARING THE SITE

Before you can do much in the way of erecting forms and pouring concrete, you have to prepare the site to receive the mix. The first step is to roughly lay out your project, and properly locate it. Then you can determine how the land lies. In some cases, this may require a bit of work with a builder's level or a transit. Drive some accurately-positioned stakes to denote the outline of your project. Indicate the perimeter of a patio or the corner points of a walkway.

Then, if the occasion warrants, set out another series of grade stakes all leveled out and with strings running between them so that as you proceed you can check for level and grade elevation at any point within the project area. Once you have established guidelines, you can remove earth from the site wherever necessary. Any sod or vegetation must be done away with, and a smooth level surface made upon which to work. Earth is removed only to the depth of the intended structure, if possible, and any necessary trenches and holes are dug for footings, columns and concrete anchors.

Wherever the concrete will be poured directly onto the earth, prepare the site as carefully as you can to avoid removing soil and having to replace it later. When possible, concrete should be poured on undisturbed native soil rather than on unstabilized fill dirt. If there are any soft spots or muck holes, these must be cleaned out and filled with sand or gravel, or a combination (often referred to as *crushed bank run* or *road base*). Any loose soil, or fill of any kind including sand and gravel, should be thoroughly tamped and compacted. Where a grade level must be brought up to fill in contour irregularities, the filling should be done with sand and gravel and not with spoil dirt removed from some other portion of the project. If the filled area extends to the sides of the forms, then it should be continued at least a foot beyond, and sloped gently away from the forms.

If the native soil has good drainage, fresh concrete can be poured directly on the ground. But if the soil is hardpan, clay, or some other type which has poor drainage, lay down a layer of either sand or gravel and pour the concrete on that. A thickness of 1 to 2 inches is sufficient. Whether the native earth has good drainage or not, if the area is subject to flooding from runoff or recurrent heavy precipitation, a minimum of 6 inches of well-compacted sand and gravel should be installed for the concrete to rest on. This will prevent water from collecting immediately underneath the slab. Where a high ground-moisture content is present around footings, drain tile is often run alongside the footing, together with a substantial amount of gravel fill, to convey away the moisture.

When the ground has been prepared and the area cleared and made ready for work, the forms can be constructed and all the other incidental details such as reinforcement and expansion joints taken care of. The tools and equipment can be made ready, the hardware laid out, and the necessary raw materials stockpiled.

Working With Concrete

Once all of the preliminary jobs of designing the project, preparing the site, building the forms and providing for reinforcement and other necessities have been completed, you can actually begin to work with the concrete itself. There are several steps involved, all of which must be taken with care if a quality product is to result. On the other hand, there is nothing difficult about any of them, complicated though they may seem at first. The four principal steps are mixing, pouring, finishing and curing.

MIXING CONCRETE

First comes what is probably the most critical part of the whole operation. This is the addition of water to the dry materials. The way not to do the job is to dump the cement, sand, and aggregate into a container, slosh in some water, and swizzle the whole mess around until it looks about right. A good cook can do this with a batch of pancakes, but no conscientious worker would make the attempt with concrete.

Water Proportions

The most crucial proportion in a concrete mix is the ratio of water—not to all the other ingredients together—but to the cement itself. Note that this amount of water includes any moisture which

115

may be contained in the sand and aggregate, as well as the water added to make the final mix. The principal reason for closely controlling the water/cement ratio is that there is an optimum point at which the cured concrete will have its greatest strength. This ratio is usually spelled out in terms of gallons of water per sack of cement (or per cubic foot; they are the same). The more water mixed in, the weaker the cement will be, and conversely, the less water used, the stronger the cement will be. Obviously there are limitations in both directions. If too little water is added, the mix will be totally unusable, no bonding will take place, and the unpourable mess will just dry out and crumble away. On the other hand, if too much water is added, the resulting mix will be a soup which is equally unworkable and valueless.

In mixing up a batch of concrete, then, there are two particular points to consider, *after* the cement/sand/aggregate proportions have been decided on, and before the water is added. One is that you must determine the approximate wetness of the sand and aggregate, as this will affect the quantity of water that you will use to make the mix. The other point is that the total quantity of water used is dependent upon the type of work that you are doing.

Determining the moisture content of the sand and aggregate is not as difficult as you might think. The aggregate seldom needs consideration, because water drains out of it readily and the material will probably be essentially dry, or at least no more than slightly damp. Unless it is obviously soaking wet, you won't have to worry about the aggregate for the home-project type of concrete work. Sand is a different matter. For a heap of sand to be completely dry would be most unusual, and there is generally a considerable quantity of moisture which must be reckoned with. This can easily amount to 1 1/2 gallons of the per-sack water requirement, or even more. Then too, the amount of sand used in various mixes will differ, which means that the moisture content will vary according to the type of mix selected.

As a matter of convenience, sand is classified in three categories: damp, wet, and very wet. The sand that you use can be placed in one of these three classes. The test that is normally used to determine the relative wetness of the sand could hardly be called scientific, but it is the accepted method and it works. Grab a handful of sand from below the surface of the pile and squeeze it tight in your hand. Then open your hand flat. If the sand is damp, it will crumble and fall apart into small clumps (Fig. 4-1A). If the sand is wet, your hand will feel damp and the sand will remain in a fist-shaped clump, with few if any of the individual grains dropping away (Fig. 4-1B). If the sand is very wet, it will compress into a mushy ball and leave your hand wet, or perhaps even dripping (Fig. 4-1C). The wetter the sand, the less water will be added to the mix.

Concrete mix formulas are often expressed in terms of the number of gallons of water to be used per sack of cement in the final mix. The number of gallons called for refers to the total amount of water in the mix, as though the sand and aggregate were bone dry. Thus, a particular mix might be called a 5-gallon mix or a 6-gallon mix. The mix may also be called a 6-gallon paste, which refers to the consistency obtained when mixing 6 gallons of water with 1 sack of dry cement. The adjustment to include the amount of moisture held within the sand is then made in the field. In theory, the range of mixes can run anywhere from a 4-gallon mix to an 8-gallon mix. In practice, however, those most commonly used run from a 5-gallon mix to a 7-gallon mix.

There is a huge number of possible combinations of water, sand, cement, and aggregate. All of the ingredient quantities are variable to a degree, and the list of possible combinations would be a mile long. However, the starting point for a given concrete mix is with one or another of the many tables which are devised for the purpose. This is called beginning "by the book." Some tables are predicated on the basic mix formula for various pastes, while others make no reference at all to the basic formula but simply list various possible mixes for different types of work. Figure 4-2 shows typical examples.

Your next step, then, is to determine an appropriate water/cement ratio and/or appropriate total mix proportions for the type of project you have undertaken by consulting the tables. After that you should test the sand for wetness. Relate the approximate wetness of the sand to the table and learn how many gallons of water per sack of cement you must mix in to make the batch. Measure the quantities of all materials as precisely as you can both in mixing the first test batch and in making up subsequent batches. This will give you the best

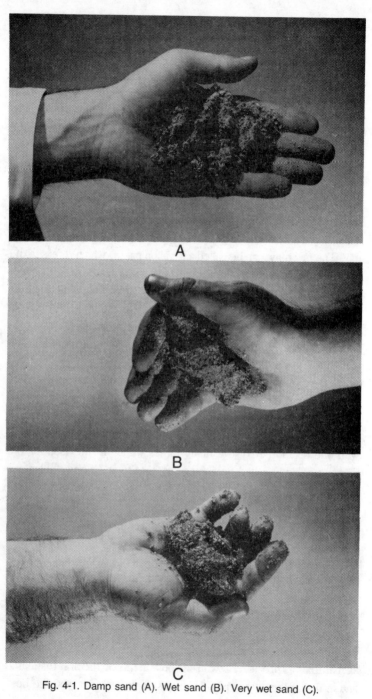

Fig. 4-1. Damp sand (A). Wet sand (B). Very wet sand (C).

possible uniformity in the finished concrete, and also serve as a stable base, a set of known factors. Furthermore, you'll have reliable base-line data for modifying the proportions if necessary.

The Mixing Process

Now, about the actual process of mixing. You have three choices: mix by hand, use a power mixer, or purchase the concrete

Paste	Mix	Water Per Sack - Gal.		
		Very Wet Sand	Wet Sand	Damp Sand
5-Gallon for:				
Severe wear				
Severe weather				
Weak acid condition				
Weak alkali condition	1:1:1 3/4	4 1/4	4 1/2	4 3/4
Severe frost action				
Topping layers				
Two- layer pours				
One-layer pours				
Industrial use	1:1 3/4:2	3 3/4	4	4 1/2
Heavy-duty condition				
6-Gallon for:				
General purpose				
Watertight				
Moderate wear				
Moderate weather	1:2 1/4:3	4 1/4	5	5 1/2
Moderate frost action				
Base layer pours				
Reinforced pours				
7-Gallon for:				
Little wear				
Little weather				
Nonwatertight				
No frost action	1:2 3/4:4	4 3/4	5 1/2	6 1/4
No exposure				
Foundations				
Footings				
Mass pours				

Fig. 4-2. Basic concrete paste mixes.

ready-mixed (which saves labor and also obviates the need for figuring proportions and laying in raw materials). More about the ready-mix situation a bit later.

Mixing by hand is a reasonable approach for batches which don't require more than 1 sack of cement. This will result in somewhere around 4 1/2 cubic feet of concrete mix, depending upon the proportions used, which means that you will have to handle somewhere around 650 pounds of material (based upon a standard concrete weight of 145 pounds per cubic foot). Once beyond the point of 1-bag mixes, then a small power mixer is much more practical than hand-mixing. When the quantity required goes beyond a couple of cubic yards, the easiest course is to purchase ready-mix concrete. The problem here is that ready-mix concrete is not available everywhere, and in outlying areas the cost of transporting ready-mix can be excessively high. Also, many suppliers have minimum quantities of 2 or perhaps 3 cubic yards of mix, or else place premium prices on any batches under 4 or 5 cubic yards.

Small batches can be mixed right in a wheelbarrow, which is a most convenient method. Wheelbarrows are commonly available in 4 1/2- to 6-cubic-foot sizes, but this capacity is when they are loaded brim-full. Obviously you won't be able to mix in them to their full capacity. From a practical standpoint, the maximum batch size for a 4 1/2-cubic-foot wheelbarrow is about 3 cubic feet, and about 4 cubic feet for a 6-cubic-foot size.

Batches of up to a cubic yard can be mixed either in a mixing tub, a hoe box, or on a mixing platform. The mixing tub shown in Fig. 1-11 is most appropriate for batches of up to 10 cubic feet or so. The hoe box is good for similar size batches. The hoe box shown in Fig. 1-12, for instance, has a maximum capacity of about 11 cubic feet, so the batch size would probably be around 8 cubic feet or less. The mixing platform (Fig. 1-13) can be made in any size you want, but 12 feet by 12 feet is a good size. This allows enough space for two men to work at the mixing at one time.

Power mixers are available in a wide range of sizes. Their capacities may be listed in terms of cubic feet absolute and cubic feet working, or in terms of bags of cement. Thus, a 1-bag mixer will handle a maximum mix predicated upon 1 bag of cement, or about 4 1/2 cubic feet. You can mix and pour a continuous series of batches, placing more concrete in a shorter time and with much less effort than by hand mixing.

When mixing by hand, start the proceedings by dumping and spreading out the required amount of sand onto the mixing surface. Place the cement on top of the sand, and spread it out. Then pile the aggregate on top. Thoroughly mix the dry materials with a square-ended shovel (or a hoe in the case of a hoe box) until the ingredients are totally combined. You shouldn't be able to see any streaks or lines of coloration, but only a uniformly-colored mass free of areas with single-ingredient concentrations. Make a depression in the center of the heap, add a small amount of water, and continue to mix. Keep this process up, without letting any of the water escape from the pile, until the entire correct amount of water has been added.

Power mixing is done a little bit differently. Start up the mixer and run in about 10 percent of the total required amount of water. Let the mixer slosh for a few seconds and then begin adding the dry materials a little bit at a time, along with the rest of the water. A shovelful of sand, a shovelful of cement, a shovelful of aggregate, a gallon of water, and so forth until the batch is complete. Don't waste any time in this process, move right along. Also, don't lose track of the amounts of the various materials that you have put into the mixer. Have the ingredients all measured out, lined up and ready to go before you begin. Effective mixing takes only a short while, and a running time of 2 or 3 minutes after you finish throwing the materials into the mixer should be more than enough. A minimum mixing time for 1 cubic yard of concrete is between 1 and 1 1/2 minutes. Never let the mixer run while you read the morning paper or wander off for a cold beer. Mix and pour out with no wasted time.

Testing the Mix

Now, after you have made the mix, how do you know if you have the right combination? You don't, but you can make some tests. None of these tests are what you might call deeply analytical, but they are surprisingly effective and accurate, especially when done by an experienced hand. The simplest test is the one most often used with small batches of concrete and is the most appropriate for the do-it-yourselfer. Follow your chosen water/cement proportions, and cement/sand/aggregate proportions, and mix up a tiny batch of

Fig. 4-3. Sample batch of concrete mix showing correct consistency.

concrete that amounts to about half a cubic foot or less. Dump the fully-mixed batch onto a level surface, and smooth off the top of the heap with a mason's trowel. If the troweled surface comes out smooth and moist, but not sopping wet, then the concrete is just about the right consistency (Fig. 4-3). If water puddles on the surface, the batch is too moist (Fig. 4-4). If the surface remains rough and pitted and will not smooth off, the mix is too dry (Fig. 4-5).

If the batch is too stiff or dry, your inclination will be to toss in a little more water, but this is exactly the wrong thing to do. Remember, that ratio between the water and the cement can be critical. Assuming that you have used the correct amount of water in the first place, you don't need more. Instead, make up a new test batch using less sand and aggregate. Eventually, you will hit the right combination. If the test batch is too wet, don't add more cement. Instead, mix in a bit more sand and aggregate until you get the right mix. In succeeding batches, return to the original quantities of sand and aggregate, but reduce the amount of water by 1 pound for every 10 pounds of sand and aggregate you had to add to make the trial batch come out right.

Fig. 4-4. This sample batch of concrete is too wet.

Fig. 4-5. This concrete batch is too dry.

Fig. 4-6. This sample batch is too sandy.

If you will be pouring a relatively large amount of concrete consisting of several successive batches, you can make a larger trial batch to begin with. This is the general rule when working with a power mixer, but you can do the same with hand-mixed batches as well. The normal trial batch is based on a 1-bag mix, but you can cut this back to a 1/2-bag mix, or even a 1/4-bag. Mix up the first batch "by the book," using the exactly correct proportions of materials and water. Then pour the batch out and check its ease of placement, stiffness, and workability.

If the batch is too sandy (Fig. 4-6), slightly reduce the amount of sand and add back the same amount of aggregate. If too stony (Fig. 4-7), reduce the amount of aggregate and replace that amount with sand. If the stiffness is too great, so that the mix does not handle and place well and is dry, reduce both the sand and the aggregate in roughly equal amounts. If the mix is soupy and sloppy, add roughly equal amounts of sand and aggregate. At first, you may find this a bit difficult to fathom, and in truth there is a great deal of judgement involved. However, after you've gained some experience, you'll be able to immediately recognize a good mix when you see one.

There is another test which is commonly used by professional concrete workers and has been found quite reliable. This is called the *slump test*. A standard steel form, sometimes called a *slump cone* (Fig. 4-8), is placed on a flat surface which will not absorb water. Concrete from the freshly mixed batch to be tested is dumped into the top of the form in three successive layers with a small scoop or garden trowel. Tamp each layer firmly, while a helper holds the form down tight to the deck so that no concrete can escape. If the mix is not stiff enough to be tamped, then it must be puddled by moving a rod up and down and around a couple of dozen times in the mix, much like churning butter. This removes air bubbles and settles the mixture into an even consistency of maximum density.

When this process is complete, the form is lifted smoothly, straight up and away from the mass of concrete. At this point one of four things will happen. If the concrete is too stiff and dry, the cone of concrete, called a slump, will crumble apart. If the mixture is stiff but nonetheless workable, the cone may simply stand there. In most cases, however, there will be sufficient moisture in the concrete for the cone to slowly settle and slump down, lowering in height and

Fig. 4-7. This sample batch is too stony.

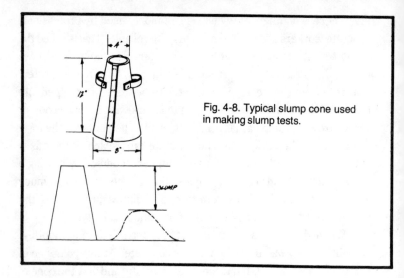

Fig. 4-8. Typical slump cone used in making slump tests.

expanding around the base. If the mix is too wet and unworkable, it will just go blah and flatten out to a pie shape.

The object of the procedure is to note the amount of slump, the distance in inches that the top of the concrete cone has settled below its original height of 12 inches. If the proportions of the mix are correct, the cone of concrete will slump gradually, stop, and retain its essential form. A normal or average slump is usually around 3 to 4 inches. For some types of work, a 6-inch slump is all right, and in a few applications even an 8-inch slump makes for a .workable mix.

The results of this test can be used in two ways. One is to arrive, by way of making several test batches, at a slump which is ideal, or at least within bounds, for a particular concrete application. Some specific parameters are shown in Fig. 4-9. The other use is to determine whether or not the given concrete mix proportions will indeed be correct for the job at hand; the slump figure serves as a double-check to make sure that everything is all right.

POURING CONCRETE

Once the initial batch has been mixed, tested and checked out, and any adjustments made as necessary, then you can go ahead and start placing the concrete. You can mix and pour as many batches of concrete as you wish or as time allows during the workday, with successive batches being poured atop one another, or side by side, or end to end. The trick is to make sure that all the pours for a given section can be completed before the material of the earliest pours begins to set up.

As long as the fresh concrete being poured upon or against is still wet, the two batches will become as one. But if the first pour is already well into the curing stage before you get back to it, you may be in trouble because the two pours will not bond together properly. This means that you will have to arrange your work schedule so that you are always pouring into fresh concrete, and so that previously poured sections do not set up to the degree that you can't get back to them to carry out the finishing operations. For some projects, staying one step ahead could involve additional manpower.

You can usually figure that you have about 45 minutes to 1 hour between the time the mixing is completed and the time when the concrete will begin to take a set. At this point, the beginning of the set-up, the concrete will become thicker and more difficult to handle. This means that you shouldn't mix more than you can pour in that

Construction	Min.	Max.
Thin walls, columns	4″	8″
Ordinary slabs	4″	8″
Furniture, thin castings	4″	8″
Heavy slabs, beams	3″	6″
Heavy walls, posts	3″	6″
Building columns	3″	5″
Reinforced slabs, beams	2″	5″
Reinforced walls	2″	5″
Reinforced walls	2″	4″
Reinforced footings	2″	4″
Slabs on ground	2″	4″
Walks, drives, patios	2″	4″
Massive sections	1″	4″
Unreinforced footings	1″	3″
Heavy pavements	1″	2″
Heavy floors	1″	2″

Fig. 4-9. Allowable slumps for various types of construction.

Fig. 4-10. Ground surface should be sprayed with water before concrete is poured.

45-minute period. But if the concrete is already in place, then you have another few minutes before the curing process has gone too far to allow bonding to a fresh batch poured on top. Curing may begin in about 1 1/2 hours.

Emplacement

The pouring itself can be done by shoveling the concrete into place, or by dumping wheelbarrow-loads into appropriate spots and then spreading the concrete. In either case, always go to the farthest point of the project and work back into the unfinished areas, so that you can stay clear of the freshly poured concrete.

With thick sections and bulk concrete, such as walls and footings, the object is to fill the forms as smoothly and rapidly as possible. Flatwork, such as sidewalks or patios, can be handled a bit differently. Here, you can set temporary plank separators between the form rails so that you can fill one small section, level it, move the separator board to the next point, fill that section, and so on. Just prior to making the pour, spray the ground surface upon which the batch of cement will rest with a mist of water from a garden hose (Fig. 4-10). The ground should be well wetted down, but be careful not to make pools. If this should happen, wait for the pools to disappear before pouring. The reason for this soaking is to insure that the relatively dry earth will not absorb an excessive amount of moisture from the lower layer of the fresh concrete mix and interfere with proper curing.

The size of the poured batches can be different than the size of the batches being mixed, and with enough men the whole affair can be kept going in a continuous process. One man can operate the mixer and make up the batches, another can ferry the fresh mix to the pouring site, while a third and perhaps a fourth as well can place, spade, and level the pour. The finishing does not normally start until about 2 to 5 hours after the first pour is made, but even so, if the pouring continues long enough, another man can be engaged in getting the finishing under way. Small projects, of course, or those that are done piecemeal over a period of several days or weeks, can be done quite handily by one man alone.

Spading

Once the first pour is completed, be that just one batch of cement or a pour of several successive batches, or as the concrete is placed, the concrete should be roughly spread and leveled (Fig. 4-11), then *spaded*. Stab a square-ended shovel blade into the fresh

Fig. 4-11. Emplacing and spreading concrete to rough level.

concrete at intervals, wiggling the blade sideways each time in order to settle the concrete and release any entrapped air. Then go over the surface of the concrete with a steel garden rake turned upside down, to approximately level the concrete with the form sides. With deep and narrow forms such as walls or footings, spade or rod the fresh concrete with a puddler to settle the mix into place and to keep large chunks of aggregate from collecting along the sides of the form. At inaccessible points, you can tap the sides of the forms with a hammer and achieve results that are nearly as good.

Screeding

After the concrete has settled out, your next step is to *strike* or *screed* it. This should be done immediately. In fact, all of these initial operations should be compressed into as short a time span as possible. The freshly mixed concrete should be dumped from the mixer into the barrows, wheeled directly into position and dumped, spread, spaded, leveled and struck, with as little waste of time as you can manage. As soon as the striking is completed, the surface should be darbied, and then there will be a breathing spell until the concrete is ripe for finishing.

The reason for this unseemly haste is a simple law of physics which says in effect that in a mixture, heavy things will sink to the bottom and light ones will rise. What happens is that as and after the concrete is handled and poured, the larger aggregate commences to settle to the bottom of the pour and water rises toward the top. With any concrete mix which does not contain air-entraining additives, this occurence, called *bleeding*, is inevitable. It is essential that the striking and darbying operations take place before bleeding occurs. If both operations are not attended to before surface water appears, the subsequent finishing processes are not likely to be successful, and the finished surface may end up being scaly, excessively dusty or powdery, crazed, or marred by other defects.

As soon as the concrete is leveled, do the screeding. Draw the *screed* or *strike-off board* across the form tops, taking the crests off the high spots on the concrete surface and filling in the troughs (Fig. 4-12). The object is to get the surface almost dead level. If there are pits or pockets left after the first pass, fill them with a small amount of fresh concrete and screed once again. Continue this until the

Fig. 4-12. Screeding a poured concrete walkway.

surface is relatively smooth and level, but make an effort to complete this activity in just two or at most three passes. Don't worry about minor imperfections in the surface: bits of aggregate sticking up, or striations left by the screed.

Darbying

Immediately after you have achieved a satisfactory surface, proceed with the *darby* (Fig. 4-13). While the screed is used with a combined downard and forward force, the darby is not. Instead, the darby should be floated gently over the surface with no weight or pressure, save for whatever is needed to iron out any obvious imperfections. The object is to embed chunks of aggregate which might be sticking up a bit, and to smooth out the slight ripples or marks left by the screed. At the same time, the surface will automatically be further leveled. Though you may find that you are scraping off a small amount of excess concrete, this should not be very much. A large amount means that you have not done the screeding job properly.

As soon as these tasks are completed, or depending upon the circumstances, even while they are going on, the pour can be started

Fig. 4-13. Darbying a freshly screeded poured concrete surface.

for the adjoining batch (Fig. 4-14). Where the sections are successive and separated by a temporary retaining board or separator, this board is removed and the excess concrete, or unformed edge where the separator had stood, is allowed to tumble down. The next load of concrete is dumped slightly in front of this edge and then shoveled into position and joined with the previous pour. The screeding then takes up just in back of the rough edge and joins the two pours into a single unbroken surface. The the darbying is carried on in the usual manner. The whole process is more or less continuous, proceeding from spot to spot and operation to operation as necessary.

Most of what has been said applies to slab work, or to any project which has a relatively large surface area to be finished. For small projects, or for those where the surfaces will be hidden or otherwise covered, such as footings, piers and walls, the process remains the same up to a point. Often the last operation on a footing, for instance, is the striking or screeding. This can be simply accomplished with a short length of scrap two-by-four run along the form top. About all that is necessary here is a relatively smooth surface, and often no further finishing is done. It is not a bad idea, however, to carry the process one step further and darby the surface as well. If nothing else, this makes for a more workmanlike job and lessens the chances for powdering or other problems. The extra step also provides a better bonding surface for a subsequent pour, as for a wall atop a footing, or for the conjoining of a masonry unit structure as the construction of a project continues.

READY-MIX

If your project requires more than one or two cubic yards of concrete, and the service is available, you may opt for buying ready-mix concrete (Fig. 4-15). In this case, you will have no need for figuring out exact proportions or laying in stocks of material, nor will you require any of the mixing equipment. But you still have to know just what kind of concrete you want, and exactly how much you will need. Or if you wish, you can accept the supplier's recommendations as to the type of concrete.

You also have to have everything absolutely ready to go when the truck appears. There are several reasons for this. One is that all that fresh concrete can't endlessly slosh around inside the revolving drum on the truck while you diddle about and try to pull things

Fig. 4-14. If enough hands are available, several operations can be carried on at once.

133

Fig. 4-15. Concrete being emplaced directly from a ready-mix truck.

together. Though the mix may contain some retardant, or may not yet have received its full complement of water (which the truck driver may add just prior to pouring), the curing process has already begun. This means that you don't have time to fool around, as just getting the mix into the forms will require a certain amount of time.

Another good reason is simple economics. Most ready-mix companies allow a certain amount of time for the driver to unload the mix at the job site. This may be as little as 15 minutes, or as much as half an hour, depending on the nature of the job and the amount of mix carried. But beyond a certain point, most suppliers start charging something on the order of 25 cents per cubic yard for every 5 minutes that the driver is delayed. If you're not ready, these charges can add up in a hurry. And if the delay is long enough, the driver will simply have to turn around and go back to the plant and dump his load, because the mix will start to set up in the truck and will become useless. You, however, will still be charged for the full load of concrete, the transportation, and the extra time spent.

As to the specific mix of concrete that you should order, this depends upon the nature of the job. Most suppliers can provide you with standard 5-, 6-, or 7-sack per yard mixes, depending upon what type of paste the job requires. In addition, you can specify the amount of water to be used. Again, this depends upon the job; some projects will need a minimum amount of water for maximum strength, while others must have a higher degree of workability and plasticity, and so, more water, at the expense of strength. Figure 4-16 will help you determine what is best for your intended project.

Have everything all prepared for the pouring well in advance of the prescheduled arrival of the first ready-mix truck. All of the forms should be complete and checked over to make sure that they are tight and sturdy and nothing is missing. The forms themselves, by the way, should be exceptionally rugged and heavily braced. They must take the weight and force of heavy concrete being poured at a rapid rate, and the forms themselves may take some bumps and bangs from the concrete chute or other equipment being used. In

FOR	ORDER
Lintels, sills	
Reinforced beams	6 1/2 sacks min.
and sections	6 gal. water max.
Reinforced floors	per cu. yd.
Top courses	
Septic tanks	
Pools & fountains	
Drives & walks	6 sacks min.
Steps & ramps	6 gal. water max.
Curbs & borders	per cu. yd.
Parking pads	
Footings	
Foundations	5 sacks min.
Retaining walls	7 gal. water max.
Substructures	per cu. yd.

Fig. 4-16. General specifications for ready-mix concrete depend upon the project.

large, thick-section pours, the rate of pouring is rapid, which means high hydrostatic pressures will be exerted on these forms, much more so than if you were pouring successive small batches yourself. Everybody will be working at top speed trying to get the job done, and there will be no time to fool around with weak or leaky forms. A collapsed form, of course, is a minor disaster, and in the case of a large pour, can be dangerous.

Make sure that the truck has plenty of room for easy access to the pouring site, as well as a spot to turn around, if necessary. The trucks are large and extremely heavy, and the driver won't thank you much if you get him mired in the muck or slide him off the edge of a loose-fill bank. If pouring chutes, chute extensions, or runways and wheelbarrows are necessary, make sure that they are all in place and set to go. Sometimes the concrete must be placed with the aid of a crane and concrete bucket, or with a concrete pumper. This means that those services will have to be coordinated to match the ready-mix truck schedule. Have all the tools and equipment at hand, including everything that you think you might possibly need.

As much of the re-bar and re-mesh as possible should already be secured in place. Reinforcing which must be added as the pouring proceeds, along with any hardware such as support plates or anchors, should be laid out in a convenient spot where they can be grabbed and quickly set in place as the work proceeds. Unless the job is a particularly small and easy one, you should not try to do everything yourself. Enlist the aid of a friend or two, or hire a couple of part-time laborers to help you out. Don't underestimate the amount of physical labor involved either; this is a tough job that requires a high level of energy in shoveling, heaving and pulling, trundling loaded wheelbarrows and such. Many a home mechanic has discovered, by way of assorted aches and pains the following morning, a whole series of muscles that he never realized he owned.

PREMIX

While ready-mix concrete is geared to the larger projects, premix or prepared mix is geared to the small ones. Premix is packaged the same way a prepared cake mix is, with all of the dry ingredients already mixed together and packed in conveniently sized bags by weight. All you have to do to complete the mix is add water and stir (Fig. 4-17).

Fig. 4-17. Small quantities of prepackaged concrete mix can be mixed in a wheelbarrow.

Using a prepared mix has a great many advantages for the do-it-yourselfer who has one or several small jobs to take care of, such as setting a couple of clothesline poles or making a single concrete step. The same material can also be used for large projects, even pools or extensive patios, if the work is to be done piecemeal and over a period of time. The unit cost of premixes is considerably higher than either ready-mix or mixing from scratch, but that is not necessarily true of the overall cost for a given job. Comparative figures should be developed for the specific project. To this you must add the convenience factor, since working with premix is easier and less problematical for home mechanics in many situations.

One of the best-known bag mixes is Sakrete. This is available not only in concrete mix form, but also in sand mix or mortar mix. The largest size for concrete is a 90-pound sack, which produces about a third of a cubic foot of concrete with a compression strength of around 4,000 pounds per square inch. Though the quantity doesn't sound like much, this is enough to pour a slab 4 feet square and 2 inches thick. The strength is more than ample for almost any home project. Other sack sizes and other brands are widely available throughout the country. Complete instructions for proper mixing are given right on the sack. This involves no more than blending the dry materials thoroughly, and then adding the designated amount of water. The resulting mix is handled and finished in exactly the same fashion as any other concrete mix.

FINISHING

Soon after the darbying has been completed, a thin film or sheen of water will appear on the concrete surface. In certain types of mixes that are particularly liquid, the bleed may amount to a thin layer or even a shallow pool of water. The finishing operation should not be started before this water has disappeared, but should be begun immediately after it does. If there seems to be an excessive amount of water on the surface, you may be forced to remove some of it with a darby or float, but the best policy is to let it evaporate.

Finishing can usually be started anywhere from 1 to 2 hours after pouring. Much depends on the weather conditions. The assumption is that the finishing can be done without having to get out onto the concrete. In the case of a large slab, the worker may actually have to move out onto the slab to do part of the finishing. It may take as much as 5 hours after pouring for the concrete to set up sufficiently to support the weight of a man. Moving out on the fresh surface is done by placing squares of plywood called *kneeling boards* at appropriate points. This distributes the worker's weight so that he will not sink into the concrete or make gouges that are difficult to repair (Fig. 4-18).

Finishing can be done either by hand or by power machines. The latter are usually reserved for extensive areas of concrete like

Fig. 4-18. Large expanses of concrete are finished from kneeling boards.

warehouse floors, and operating the machines does take some training and experience. Hand finishing works as well, however, and in many instances better, even though the process may take a bit more time. With either method, the key to a good finish is to not overdo the operation. Whatever procedure is being used should be carried out as rapidly and efficiently as possible, with an absolute minimum of going back over portions that have already been done. This is because repeated passes over the surface of the concrete with finishing tools bring another layer of water to the surface, and the bleeding begins all over again. The result after the concrete has cured is the same as though striking or darbying were done after the appearance of bleed water. You won't get a good finish, and you will get a lot of powdering.

Floating

Probably the most common finishing method of all is *floating* (Fig. 4-19). The appearance and texture of the finished concrete surface depends upon the material from which the float is made. A wood float will impart a slightly rough surface, which is ideal for driveways, patios, and walks. A metal float, on the other hand, will produce a smooth surface. The hand float is the easiest to use,

Fig. 4-19. Finishing concrete surface with magnesium hand float.

provided that you can reach the entire work area and that the surface to be finished is not overly large. The long-handled float works best for the larger areas and those that cannot be easily reached. This type of float will finish a greater area in a shorter time, primarily because the float itself is bigger and you can work more rapidly from an upright position (Fig. 4-20). The extra-long-handled float, or bull float, is used on large slabs, and is also occasionally used for darbying as well. Floating can also be done from points out on the fresh concrete, provided kneeling boards are used.

An old float works better than a new one because the sharp edges have been worn away and the tool will not catch in the concrete and create gouges. If your float is brand new, it would be helpful to take a file or a plane to it to round off any sharp corners and all of the bottom edges. Floating is an easy process to master, and you'll soon get the hang of it. Put almost no pressure or weight upon the float; guide it smoothly back and forth and around and about, letting the slight weight of the float itself and the sliding motion do the work. Tilt the blade up slightly, toward the direction of travel. The only time any pressure is needed is when you encounter a slight bump that needs to be ironed out. Even then, you should proceed with caution. Start floating at the farthest end of the concrete surface, which should also be the area that was poured first, and work backward until the job is done. Any excess moisture or particles that you might bring up should be floated off the sides of the form. This is your last chance of getting a perfectly smooth surface, too, so it pays to keep a close visual check on the finished surface as you move along.

Steel Troweling

Another finishing process which is sometimes used is *steel troweling* (Fig. 4-21). This process is actually little different than floating, and the steel trowel is similar to a metal hand float with a springy blade. When properly done, this process produces an ex-

Fig. 4-20. Slab being floated with a bull-float.

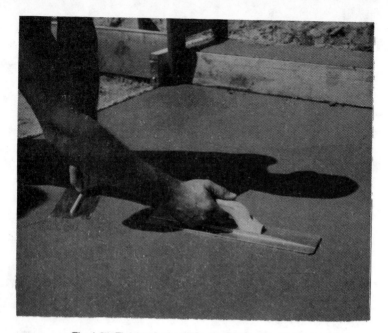

Fig. 4-21. Finishing concrete surface with a steel trowel.

tremely smooth finish without a ripple or groove. When steel-troweled concrete gets wet, however, it can quickly become as slick as greased glass; and a little fresh snow on top can make it lethal. This is a dangerous finish to use on steps and patios or wherever people will be walking about, so have a care about where you use it.

The job is done in a manner somewhat similar to ordinary steel floating, but there are a few important differences. Troweling begins shortly after the floating process has been completed. It is a good idea to let a few minutes elapse before beginning, so that the surface can firm up a bit. The worker then goes over the entire surface with a large steel hand trowel, usually with side-to-side or back-to-front strokes that are firm and even. Unlike floating, troweling requires that the blade be kept as flat as possible; if it is tilted, a ripply effect will result. A fair amount of pressure is also put on the trowel, but not enough to make the blade dig in and gouge. The pressure is always applied evenly over the surface of the trowel and remains the same from stroke to stroke. As little reworking as possible should be done.

Often the finishing process is terminated at this point, and a tough, smooth and hard surface results. However, it has not yet approached the glassy stage. The surface can be made even smoother by subsequent trowelings. After the first pass, wait for a few moments before starting the second to allow the surface to firm up even more. Then go back and do the whole job over again. Subsequent trowelings should also be done with short time intervals, and with progressively smaller trowels so that plenty of pressure can be applied to attain a super-smooth surface. After 3 or 4 passes, the surface will be so smooth that after curing it may even have a slight sheen.

Brooming

The opposite of a steel-troweled finish is the *broomed* finish, which results in a rough but evenly textured surface. This too is done after the floating process. It's a simple enough job, but requires a steady arm and a good eye. As you may have guessed, the tool used is a broom, usually some sort of floor broom (Fig. 4-22).

Fig. 4-22. Brooming a fresh concrete slab to provide a textured effect.

Fig. 4-23. Brooming with a coarse brush creates deep striations and an emphatic texture.

To broom a fresh concrete surface, just pull the broom toward you across the concrete in a straight line, exerting no pressure upon the broom head. Overlap each stroke just a bit. When the surface is a large one, you will have to do the brooming in several segments, using kneeling boards and shifting around as necessary. The trick here is to set the broom gently and accurately into previously broomed sections in a way that avoids lap marks. This requires a nice touch, but can be done. An alternative possibility is to broom in sections, and then refloat over the lap marks in certain areas, creating a pattern of broomed sections alternating with faintly lined sections. In some cases, you may be able to tape a long extension pole to the broom handle and get by this way.

The texture of the surface can be controlled by using finer or coarser bristles or stiffer or more resilient brushes, as you wish (Fig. 4-23). You can test the effect of different brooms on a freshly floated surface and then refloat preparatory to the final brooming, provided that you do not over-finish any given section of concrete in the process. Whatever the texture, the brooming pattern is best run at right angles to the normal traffic pattern, as across a walkway or driveway. This is also an ideal way to roughen step surfaces for safety.

Grooving

Grooving is usually done to walkways, and often to driveways as well. This should be done after the floating has been completed. Lay a plank across the concrete to serve as a straightedge, and guide the groover along the plank edge, using enough pressure to cut the groove to full depth, but not so much that the flat portion of the groover gouges the surface of the concrete (Fig. 4-24). If undesirable, the smooth mark left by the flat portions of the groover's track can be erased by refloating. Be careful, though, not to fill the groove you have just cut. If the final finish is a broomed one, the grooving can be done either before or after the brooming, depending upon whether or not the smooth strips left by the groover are wanted.

Edging

Edging is normally done to walks and driveways, and sometimes patios as well. It is done in the same manner and at the same time as the grooving, by running an edger along between the concrete and the form rail (Fig. 4-25). This results in a slightly rounded edge, as well as a smooth border. The border can be either left as is

Fig. 4-24. Grooving a freshly poured sidewalk.

Fig. 4-25. Finishing the perimeter of a patio slab with an edging tool.

or eliminated, if you wish, by floating or brooming. Rounding of the edge isn't only done for the sake of appearance, it's also done to make the edge less prone to chip or crumble away.

Finishing Air-Entrained Concrete

All that has gone before refers to the various standard types of concrete mixes. However, if you are working with an air-entrained concrete, the situation is different. This is because the consistency of the concrete is not quite the same and the millions of tiny air bubbles throughout the mix tend to hold everything in relatively stable suspension. Also, less water is used in making up the mix. These two facts mean that little or no bleeding takes place and floating can start much sooner. The recommended floats are made from either magnesium or aluminum; they're light and will not drag as a wood float will. Other than that, the finishing processes are just the same, except that you must take care not to get caught short by waiting too long. This type of concrete sets up and begins to cure rather quickly, especially when the weather is warm and dry. On the other hand, you can also expect to get a better finish with less work than you would with standard concrete mixes.

Fancy Finishes

Standard finishing techniques are employed entirely to give the concrete a smooth, level, dense and weatherable surface; the appearance is rather bland. But with a little extra work you can provide decorative finishes which are unusual and appealing. In fact, artistry and imagination can do much to enhance the final appearance of poured concrete.

Texturing is one popular approach to decorative finishes, and the many variations can be done with ease. The surface shown in Fig. 4-26 is a swirl done with a wood float. After the surface is darbied, it is gone over with a float, held flat and worked in regular semicircular sweeps. The pattern can be changed to waves, circles, or any other form that appeals, just by manipulating the float. A coarse texture is made with a wood float, a medium one with a metal float, and a fine one with a steel trowel. Texturing can also be done with a broom, as shown in Fig. 4-27. Here the pattern is controlled

Fig. 4-26. Swirl texture on concrete surface made with float.

Fig. 4-27. Wavy broom texture on concrete surface could also be made with a float.

by the broom movement, and the texture by the relative stiffness and coarseness of the bristles. Alternating or varied textures can be made on the same surface to create greater diversity.

The *travertine* texture (Fig. 4-28) is one that can be used for an interesting effect. The concrete is first darbied in the usual fashion. This is followed by brooming with a stiff and fairly coarse-bristled broom to provide keying for the coating to follow. The finish coat is mortar, in the ratio of 1 sack of white portland cement to 2 cubic feet of sand, mixed to a consistency of a thick paint. A mineral pigment may be added, in the amount of 1/4 pound per sack of cement. Yellow is a favorite, because it provides a pleasing effect similar to that of travertine marble, but any color can be used.

This dash coat of mortar is thrown, literally, onto the fresh concrete surface by flicking the mix off a stiff dash brush in gobs, making an uneven and irregular surface full of lumps and bumps, but occupying every sector of the pour. Minimum thickness of the coating should be about 1/4-inch, maximum about 1/2-inch. The dash coat is left alone until it has cured sufficiently to get out on with kneeling boards without making impressions; constant checking of the curing progress is necessary so that finishing can start at the right time. Then the coating is steel troweled, to knock off the tops of the bumps and pull some of the material into the low spots. At the same time, variations in the textures can be made to form decorative patterns, or the surface can be scored or otherwise treated for further effect.

Dividing is an easy method of providing a decorative effect, and has the added advantage to the do-it-yourselfer of breaking the job up into small sections that can be easily poured piecemeal as spare time permits. Dividing consists of breaking a large slab up into sections by installing wood divider strips and sides in whatever geometric pattern appeals or is suitable for the project at hand. The strips are nailed together to make a rigid structure, usually of redwood, cypress, or some other treated wood. Whatever the wood, it should be made resistant to rot and insects, and sealed with a clear sealer to improve weatherability. The strips make up the forms, and the tops are covered with masking tape before pouring to protect them (Fig. 4-29). Pouring and finishing of the concrete are done in the usual manner, followed by removal of the tape and whatever cleanup is necessary. When the weather is hot and sunny, peel that tape off as soon as possible, before the stickum turns to goo.

Fig. 4-28. Travertine texture effect.

Fig. 4-29. Apply masking tape to wood dividers for protection during concrete pouring and finishing.

Pattern finishes have always been popular, and are easy to do. One method is shown in Fig. 4-30. After the surface has been darbied and the moisture sheen has disappeared, tool random lines into the concrete. Use a jointing tool, or just a piece of bent pipe. After the surface sets up further, go over it with a float, and again run the jointing tool along the lines to smooth them up. Probably a second go-round with the float will be necessary to tidy everything up, followed by a careful touchup with the jointing tool. Crumbs can be brushed away with an old, soft-bristled paintbrush.

Another possibility is to set *joint forms* in place. The forms can be made from 1/4-inch-thick lattice stock, randomly scalloped as

Fig. 4-30. Random joints can be tooled into a concrete slab with a jointing tool or a piece of bent pipe.

shown in Fig. 4-31. The bottom edges should be rounded so that they can be easily removed from the concrete later. After darbying the concrete surface, but somewhat before the usual time for finishing to start, soak the wood strips in water and lay them out on the

Fig. 4-31. Small wood joint forms carved from lattice strip.

Fig. 4-32. Joint forms laid out in pattern on fresh concrete surface.

concrete (Fig. 4-32). After the pattern is established, push the strips down into the concrete with a float (Fig. 4-33), and float the entire surface so that everything is level. Clean any excess concrete or cement paste from the tops of the strips with a small trowel (Fig.

Fig. 4-33. Joint forms being pushed down into fresh concrete, level with surface.

Fig. 4-34. Excess paste being cleaned off joint forms after floating.

4-34), and then proceed with further floating, steel troweling or brooming in the normal manner. Leave the strips in place for 24 hours, and then pry them up carefully with a trowel tip.

The impressed patterns are usually partly filled with a mortar of contrasting color for heightened effect (Fig. 4-35). Flood the slab

Fig. 4-35. Joint indentations left after form removal being filled with colored mortar.

153

Fig. 4-36. Leaves being pressed carefully into freshly finished concrete surface.

with water first, and let it dry until the surface is merely moist. Brush a paste of cement and water, with the consistency of thick paint, into the joints and immediately follow up with the mortar. Take care not to smear either the mortar or the paste onto the slab, and have sponge and a bucket of clear water handy so that you can clean up as you go along. The mortar should be packed in firmly and smoothly, with the top surface lying below the concrete surface by a fraction of an inch.

A *fossil* effect is made in similar fashion. Figure 4-36 shows leaves being impressed into a freshly finished surface, but other objects could be chosen a substitutes. The objects should be pressed firmly into the surface, deep enough so that they can be troweled over smoothly, but not so deep that cement paste accumulates over them. After the cement is sufficiently cured to be good and firm, the objects are carefully removed. Patterns can also be impressed by twisting into the surface different-sized tin cans (Fig. 4-37), cookie cutters, or anything else that comes to mind.

Texturing with rock salt is an easy job, but is not recommended for climates where a lot of freezing goes on in the winter. After all the finish work is done to the slab, rock salt is scattered around on the

154

surface, and then pushed down with a float or a roller, without disturbing the finish. The salt pellets should stick up just slightly above the surface, and need not be surrounded with a cement paste. After the concrete has cured, the surface is flooded with water, dislodging and dissolving the salt and leaving tiny craters and pits (Fig. 4-38).

There are a great many different kinds of *exposed-aggregate* finishes, and there are several methods of making them. One way is to pour a fairly stiff concrete with maximum 3/4-inch aggregate into the forms, striking and darbying as usual but making the slab surface about 1/2 inch lower than the finished grade level. A previously selected, washed aggregate of small colored stones, bits of marble

Fig. 4-37. Pattern finish being made by impressing tin cans of various sizes.

Fig. 4-38. Rock salt texture finish on either side of a pebble-embedded feature strip.

or other decorative material is shoveled onto the fresh surface in an even layer (Fig. 4-39). The layer is tapped and pushed carefully down into the concrete with a darby or hand float (Fig. 4-40), further embedded with a bullfloat, and finally worked completely down out of

Fig. 4-39. Throwing selected and graded clean aggregate onto a fresh concrete slab.

Fig. 4-40. Leveling and bedding the aggregate into the fresh concrete.

sight with a hand float (Fig. 4-41). The layer should remain even, and be covered by about a 1/8-inch layer of cement paste. When the process is complete, the slab will look much like any other. Now the slab must set up enough so that a man on a kneeling board will leave no impression on the surface. Brush the surface with a stiff-bristled nylon broom to remove as much of the surface mortar as possible,

Fig. 4-41. Floating the concrete surface to just cover the aggregate.

157

Fig. 4-42. Spraying and scrubbing the concrete surface to expose the aggregate. The two operations can also be done separately with hose and brush, or by two men working together.

but without dislodging any of the aggregate. Next comes washing with a fine spray of water and brushing with a stiff-bristled brush at the same time, until the aggregate is evenly exposed but remains well bedded (Fig. 4-42).

Another method calls for hand-placing larger preselected stones, like cobblestones, deep into the concrete. This can be at random, in regular rows or strips, or in patterns of your choice. The remainder of the process is about the same as previously described for the small aggregate, and results in a finish like the one in Fig. 4-43.

CURING

Proper curing is an important part of any concrete job. Sometimes after the finishing has been completed the workers simply walk away from the job and leave the whole business until they come back to strip the forms a few days later. Frequently luck holds, and the concrete does cure well enough by itself. Often, however, there is a serious loss in strength because of improper hydration, to the point where only half and at times even less of the potential strength of the concrete is realized. Trouble is, curing can sometimes be a long and involved proposition, and many folks just don't want to bother. The problem stems from the evaporation of the water from the concrete surfaces, which reduces hydration, which in turn diminishes strength. In severe cases, this can lead to surfaces that are easily damaged, that powder and chalk, crack and crumble, and heaven knows what-all else.

We are faced here with a definite disparity between the ideal and the practical. Theoretically the period of hydration can extend indefinitely, but in practice the process is all but over by the end of the first month. Theoretically, the ideal strength of a given mix of concrete can be achieved by *moist-curing* for an indefinite period of time. Practically, nobody has the time or inclination to fool around for any extensive length of time. And practically, concrete which is moist-cured for only a short period will achieve anywhere from 75 percent to about 90 percent of its ideal or potential strength, which is usually more than ample for the job involved.

Other practicalities also interfere. We can't do much about the weather, can't even combat it very effectively, and this plays a definite part. The specifications of some types of mixtures, or structures, do indeed call for lengthy curing periods involving a lot of work, but many others do not. As a general rule, you can figure that a curing period of 5 to 7 days will afford the maximum practical results. In the case of slabs poured directly on the ground in ideal (70°)

Fig. 4-43. Driveway of embedded and exposed cobblestones.

Fig. 4-44. Curing concrete slab with wet burlap.

temperatures with little or no wind, with high humidity and protection from the sun, the curing time can be as little as 2 1/2 to 4 days.

There is nothing difficult or strenuous about the curing process; it just has to be kept after all the time. And whatever method is used, the curing process should begin directly after formed concrete is screeded or slab concrete is finished.

If the concrete was poured in steel forms, just leave the forms in place as long as you can. Keep the exposed top surface of the concrete continuously wet, either by sprinkling or by laying out a canvas soaker-type hose. If the forms are wood, they should be kept wet by spraying them with a hose. Again, the top surface should be kept wet also. One of the best methods for curing slab concrete is to sprinkle it with water. Don't direct a hard stream right on the new surface, but rather provide a fine mist of water continuously from a garden hose. If you have to, use several hoses appropriately positioned. This system has to be constantly checked to make sure neither too little nor too much water is being used, or that the wind isn't blowing the spray away and at the same time drying the surface.

Intermittent sprinkling is helpful in curing, but usually is not too good an idea unless the weather happens to be calm, cloudy, and humid and the slab can be kept perpetually moist. If allowed to become alternately wet and dry, you may run into problems in the way of cracking or other defects in the finish. An alternative method is to spread burlap, old blankets, or mats over the concrete and keep them wet (Fig. 4-44). This requires only periodic sprinkling, and serves as a sun shield and confers protection from wind. Straw and sometimes hay can also be used, and should be liberally sprinkled often. The problem with this material is that it can dry out rapidly and blow away. Unless spread quite thick, it does not give complete protection. Straw is pretty expensive nowadays, too.

Another commonly used method of curing is to cover the slab or the form top with sheeting (Fig. 4-45). You can buy waterproof paper for just this purpose which provides excellent protection from sun and wind and practically eliminates evaporation from the concrete surface. The paper has to be handled carefully, rolled out evenly, and weighted so that it will not blow away. The cost is rather high, but the paper is reusable if taken care of.

Polyethylene sheeting, often called "construction plastic," is commonly used for this purpose. It is generally of the clear variety, though clear is the least effective type. The plastic should be pig-

Fig. 4-45. Moist-curing concrete slab by covering with plastic sheet.

Fig. 4-46. Spraying on curing compound. Colored dye in the compound helps you to get an even coat.

mented for best results. Black plastic can be used in cloudy, cool weather, but not in hot and sunny weather. White plastic can be used under any conditions, and serves as an excellent reflector for hot sun. Plastic has the added advantages of being inexpensive, easy to handle, and impervious to moisture. Just remember to handle it carefully so that it does not tear or puncture, and weight the edges so that it cannot blow away and so that wind cannot get underneath.

Another good method is to use a special *concrete-curing compound* which is sprayed on with either a hand sprayer or a power sprayer as soon as possible after the finishing is done (Fig. 4-46). There are several types of these compounds, both dark and light in color. When sprayed on, the compound forms a skin over the fresh concrete which halts the evaporation of moisture. A dark compound can be used in cool weather, and will absorb the sun's heat as an aid to curing. In hot weather, a white compound can be used to reflect some of the sun's heat. This reduces the concrete's temperature and slows the hydration process to a more effective rate. For average conditions, you can use a clear compound. Some of the clear compounds are sold with an accompanying colored dye so that you can tell whether or not you have completely covered the concrete surface. The dye becomes invisible after a short time. Compounds cost little, are easy to apply, and can be sprayed onto structures of any shape or configuration. As long as the coverage is thorough and complete, compounds do an excellent job. They can also be applied after a period of moist-curing to easily extend the total curing time to many days without interfering with the progress of further work on or around the project. Compounds also work nicely on formed concrete where the forms have to be stripped away before the critical early-curing period has been completed.

STRIPPING

Removal of the forms is generally referred to as *stripping*. For most types of concrete work, form stripping is usually done from 3 to 5 days after the pour is made to give the concrete as much time as possible to cure with a minimum of extra work. Frequently it is necessary to strip forms earlier than this, sometimes within 2 days, and sometimes, as in the case of high-early-strength concrete, in as little as 1 day. However, forms should always be left in place as long as is practicable.

The one exception to this occurs when the forms are wood and the weather is hot, windy, and dry, and there is no reasonable means available to keep the wood forms soaked with water. In this case, the forms should be removed soon after the concrete becomes solid and self-supporting, even though the curing process has not progressed very far. Then the exposed concrete areas must be covered with sheeting, or preferably with curing compound, so that proper curing can take place.

If you can keep the forms in place for a longer period of time without creating any problems, this is all to the good, especially if work is continuing around the job site. Not only will the concrete cure better, assuming that exposed areas and wood forms are kept properly moist, but also the green concrete will be well protected from mechanical damage. If the forms are removed within a few days, then the use of moisture treatment or curing compounds on exposed surfaces for another few days or even a couple of weeks certainly will do no harm and may well help the curing process considerably. This is by no means mandatory, and in fact is seldom done, especially on small projects.

Stripping requires a modicum of caution and commonsense. Some workmen seem to take great delight in attacking a set of wood forms as though they were razing an old building. But the best approach by far is to disassemble the forms methodically and carefully, starting at the outermost points and working in. After all, 95 percent of that expensive form material can be used again, either for more forms or for some other purpose. Also, at this stage of curing, concrete is quite susceptible to mechanical damage. All of the components of the form which lie in contact with the concrete should be removed with special care.

CLEANUP

In the interest of safety, efficiency, and convenience on and around the job site, cleanup should begin either as the stripping process goes on, or immediately afterwards.

Protruding nails should be pulled from the form boards, and all forming material stacked and out of the way. Equipment and tools can be cleaned and stored, and all of the general debris that seems to accumulate during any such job raked and shoveled up and disposed of. This is a good time, too, to check over the fresh concrete and make sure that everything is all right. Clip off any protruding tie wires, and clean out the threads of anchor bolts with a wire brush. If there are nibs or protrusions of concrete on the formed surfaces resulting from cracks or defects in the form sides, this is a good time to get rid of them, while the concrete is still relatively soft. Larger chunks can be chipped away gently with a cold chisel and hammer, while smaller ones can be scrubbed off with a piece of brick or concrete block, just as though you were using sandpaper.

BAD WEATHER POURING

As mentioned earlier, pouring concrete in bad weather entails a special set of conditions and procedures which are necessary to ensure maximum strength and a good finished job. First, what does bad weather mean, as far as pouring concrete is concerned? Ideal conditions for pouring are temperatures of about 60° to 70°, little or no wind, medium to high humidity and little or no sun, or at least a weak sun which provides little noticeable surface heat. Bad weather, then, is everything that's left.

Any conditions which either alone or in combination will produce high temperatures and rapid drying mean that special precautions must be taken. Hot sun, temperatures over 80°, low humidity, strong breezes or wind, all can spell trouble. In arid areas of the Southwest and West these conditions are encountered more often than not, and even 60° with strong wind and a humidity of 15 to 20 percent can cause a great many difficulties.

When pouring where such conditions prevail, you can take steps to avoid or minimize the attendant problems. One possibility is to do the pouring in the evening or at night, giving the concrete a chance to begin curing in cooler temperatures and while the wind is down. Where quick evaporation is a problem, you can use more water in the concrete mix. The cement/water ratio should remain constant, however, and this in turn means less sand and aggregate and more cement, at the same time making a mix of proper slump or plasticity.

Cooling of the materials is also possible. Keep everything protected from the sun. Piles of aggregate should be soaked with a hose and covered with a tarp, which also should be kept soaked. Sand piles should not be soaked, but just covered. Make the mix with cold water. Add plenty of ice if the water supply is warm, to bring the temperature of the water down to around 35° or so. Just before pouring, soak the forms and the ground with cold water, especially if they are in the hot sun.

Set up windbreaks of plywood sheets, tarps, plastic sheeting or anything else that happens to be handy, to help reduce rapid drying. If the wind isn't blowing but might start before you can complete the pouring and finishing, set up the windbreaks anyway—you won't have time during the job. Sunbreaks are most helpful, too, if you can arrange them; tarps or white polyethylene sheets work well. Adding hydration retardants to the concrete mix is also a good idea, especially if you are sure the temperatures will reach into the high 80s or more. If you are working from a ready-mix truck, be set to pour immediately after the truck arrives, so the churning mix doesn't have a chance to sit out in the hot sun. Get to the finishing as soon as you can, even if that means hiring on extra hands, and begin moist-curing as soon as the finishing is done. Continue this for at least 24 hours, longer if you can, and follow up with a curing compound or plastic sheeting. In short, do everything you can think of to lessen evaporation, reduce drying and hardening, and keep concrete temperature down.

At the opposite end of the scale, the problem is not with drying, but with cold temperatures. If the temperature is, or expected to be at any time during the pouring and initial curing stages, 40° or lower, a good many precautions must be taken. As the temperature of the concrete itself falls below about 50° (a bit more for thin sections, a bit less for massive sections), the curing action slows considerably. The closer the temperatures get to 32°, the greater becomes the danger of freezing, which will destroy the strength of the concrete.

There are plenty of methods that professional concrete workers employ so that they can keep pouring no matter how cold the weather, but for the most part these involve a lot of extra work and expense. Professionals build shelters and protective covers, add calcium chloride to the mix, heat the materials, mix with hot water, heat the forms and/or the ground, install heaters and use special cements. For the do-it-yourselfer, the easiest way out is to forget about working with concrete when the temperatures are expected to be less than 45° or so. If you do get caught short by a sudden cold snap wherein there is a danger of the concrete freezing at any time during the first 3 or 4 days, jury-rig a protective cover of some sort with boards, tarps, plastic or whatever. If necessary, set up a heat source. A string of light bulbs works well, or an electric space heater, even an electric blanket. If need be, you can rent an oil-fired hot air blower from an equipment rental house. Whatever you use, all that is necessary is to keep the concrete surfaces a few degrees above the freezing mark.

Mortar, Grout and Stucco

Though concrete is probably the most-used of the materials blended with a cement base, there are three others which are common in outdoor home construction projects. These are mortar, grout and stucco. Of the three, mortar is both the most important and the most widely used.

MORTAR

Mortar is the binding agent which ties together the masonry units in building-block construction. These units may be brick, concrete block, tile, stone, etc. The way in which mortar is mixed and applied is extremely important to this type of construction, because the mortar joints are by far the weakest part of the completed structure. Correct proportions, proper mixing, and top-grade workmanship are needed to produce a strong and high-quality finished product. Deviations from the proper mortar formulas and mixes, especially when combined with sloppy workmanship and poorly laid masonry units, will result in an unsatisfactory structure which may be short-lived or even dangerous.

Along with its principal function as a binder, the mortar must also serve as a weatherproofer, preventing moisture and air currents from entering a structure by insuring an unbroken facade of complete integrity. Also, on many occasions the mortar and the particular method of tooling the joints, as well as the jointing pattern, may be part and parcel of the overall decorative and architectural effects of the structure.

Mortar Types

The ingredients used in any mortar mix are cement, hydrated lime, and sand—plus enough water to make a suitable mix. Three kinds of cement can be used. The first is the standard ASTM Type I portland cement, the same as is used in concrete mixes. The second kind is a pure white portland cement which is always used with a coloring agent. It is utilized when the builder wishes to create unobtrusive mortar joints which harmonize with the hue of the masonry units. Alternatively, it is used to produce colored mortars which contrast with the hue of the masonry unit and thereby enhance the overall appearance of the structure. The third kind of cement is a special mix called *masonry cement*, which already contains a certain amount of lime.

Mortar can be mixed from either portland cement or masonry cement, and some recipes call for a combination of the two. White portland, however, is always used alone, since the addition of gray masonry cement would cancel the desired whiteness of the base mix. Further, the masonry cement is itself available in Type I *and* Type II, of which the latter is most commonly used. All of these cements are purchased by the sack (1 cubic foot of material).

The single aggregate used in mortar mixes is sand, and not just any old sand will do. For best results, the sand should be sparkling clean, and contain not a bit of debris, dust or foreign material. It should be thoroughly washed. The grading is also important, and there should be a range of grains running from a maximum size of about 1/8-inch diameter all the way down to tiny particles, with a relatively even span of gradations in between. If all of the particles were tiny, the resulting soup would be hard to work with, and if all the particles were large, neither the bonding nor the workability would be good. But principally, a well-graded sand ensures that there will be only the barest minimum of air voids, since those would be filled by the multisized particles. Consequently, less cement is needed to fill the few remaining voids. This makes for strong bonding and an economical mix. The sand particles themselves should be rough, multifaced, and sharp-edged. When you order sand for mortar mixes, specify washed, sharp mortar sand, and you'll have the correct variety.

Hydrated lime is the ingredient that makes mortar what it is. Without the lime, you would have just a porridge of straight-sand concrete which would be virtually unworkable as a binder for masonry units unless it was very rich in cement, though the mix could be poured in place and treated as a concrete. The lime, even though it diminishes the potential strength of the mixture, provides the necessary workability and plasticity that a good mortar must have. By using the proper amount of lime for a given mix, the mortar can be effectively worked and the strength of the cured joints will not be significantly reduced.

Lime is also bought by the sack, and added to the dry mix as called for. Incidentally, lime also has a nasty habit of eating holes in your skin. The dry powder is inert, but as soon as moisture is added, including perspiration from the skin, the action begins. Working barehanded with wet mortar is not a good practice, because the lime can burn away deep patches of skin in a short period of time, and the result can be painful. Wear gloves, and wash your hands frequently in fresh water. The same advice holds true for wet concrete mixes—since lime is also present there—but to a lesser degree.

The last ingredient in the mix is water. As in concrete mixes, the water should be clean and relatively pure. The fewer minerals and foreign ingredients the better. If the water is drinkable, it can be used to mix mortar. Under no circumstances should salt or alkaline water be used. Cool or cold tap water is fine.

The different types of mortar mixes that are in most common use today are designated by letters of the alphabet. There are five categories intended for numerous different kinds of work, and there is considerable overlapping among the types. In all cases, the prescribed proportions should be adhered to closely.

Type M mortar is made up of 1 part portland cement, 1/4 part hydrated lime, and 3 parts sand. The same type may be mixed by using 1 part portland cement, 1 part masonry cement Type II, and 6 parts sand. This is a general-use mortar, and can be used for just about any kind of job. In particular, it is the best of all for below-grade situations, such as for setting the first course or two of masonry units on a concrete footing, where the joints will later be covered by earth. This is also the mix to use for mortaring up retaining walls, when one side of the wall will be embedded in the earth, or for joining the elements of a walkway when the same situation prevails.

Type S mortar is mixed with 1 part portland cement, 1/2 part hydrated lime, and 4 1/2 parts sand. The alternative mix for the same type uses 1/2 part portland cement, 1 part masonry cement Type II, and 4 1/2 parts sand. This too is a general-purpose mortar which can be used for just about any project you might conjure up. It is distinguished for its strong resistance to lateral pressure. This is the kind of pressure which can be expected against a tall retaining wall, for instance, or against the walls of a cellar entrance bulkhead constructed from masonry units.

Type N mortar is mixed with 1 part portland cement, 1 part hydrated lime, and 6 parts sand. The same type can be made from 1 part masonry cement Type II, and 3 parts sand. Again, this is a general-purpose mortar, but not quite as broadly applicable as the previous two. In particular, it is used for constructing foundation walls made from solid masonry units, and is suitable for nearly all general usages in above-grade situations. It is excellent for structures which will be exposed to heavy weathering and severe weather conditions such as can be experienced in coastal areas or in some high mountain sections.

Type O mortar is made by mixing 1 part portland cement, 2 parts hydrated lime, and 9 parts sand. The same mix can be formulated with 1 part of either Type I or Type II masonry cement and 3 parts of sand. This mix is much less utilitarian than the previous ones, and is primarily used in putting together load-bearing walls of solid masonry units in above-grade locations where freezing and thawing is not a problem and the compressive loads are less than 100 pounds per square inch. It is not satisfactory for rubblestone or fieldstone work, even though those materials are sometimes considered to be solid masonry units.

The fifth and last kind of mortar, Type K, is made by mixing 1 part portland cement, 3 parts hydrated lime, and 12 parts sand. Note the extremely high proportion of sand to the rest of the materials. This is a limited-use mortar, primarily for large solid masonry walls which have considerable support. This mix would not normally be used for home projects.

There is also a more general approach to determining the kind of mix you should use. The Portland Cement Association looks at the situation this way. For general-purpose uses and in anticipation of only ordinary conditions, you can make your mix with 1 part portland cement, 1 to 1 1/4 parts hydrated lime, and 4 to 6 parts sand. Or, you can use 1 part of ASTM Specification C91 masonry cement, and 2 to 3 parts sand. Note that this mix is similar to Types S and N, but isn't quite one or the other.

However, if you anticipate conditions which are more severe than average (whatever that may be), then you choose a stouter mortar. Determining such conditions is a matter of judgment. Extra heavy loads would be a case in point, and so would be chronically adverse weather, high or freakish winds, or severe frost action (as is prevalent in the northeastern United States). Under such circumstances, the recommended mix is 1 part ASTM Specification C91 masonry cement, 1 part portland cement, and 4 to 6 parts sand. Or, you can use 1 part portland cement, 0 to 1/4 part hydrated lime, and 2 to 3 parts sand. Note that this mix is approximately a Type M, though some latitude is allowed in the quantity of sand used.

There are also many "in-between" mixes (Fig. 5-1). In all of these mortar mixes, the higher the percentage of portland cement the greater the compression strength of the cured mortar in the joints, though other factors such as the percentage of hydrated lime also play a part. But by and large, the Type M mortar or richer will provide the greatest compression strength, while the Type K provides the least. It is practically impossible to tell exactly what the tensile or compression strengths of any given batch of mortar will be, because there are so many variables involved and no definitive way of finding an answer. Pinpoint figures can only be established by working under controlled conditions and with laboratory tests. However, Type M mortar can have a compression strength of around 5000 pounds per square inch, while Type S would be expected to have about half that. Type N mortar has a compression strength of 1000 pounds per square inch or less, about half that of Type S. And so on down the line.

Where only small batches of mortar are needed, or where a larger job will be done a little bit at a time, the do-it-yourselfer's best approach is to use premixed mortar. All of the dry ingredients are

Mix Ratio	Cement Sacks	Lime Lbs.	Sand Cu. Yd.
1:1/10:2	13.00	52	0.96
1:1/10:3	9.00	36	1.00
1:1/10:4	6.75	27	1.00
1:1/4:2	12.70	127	0.94
1:1/4:3	9.00	90	1.00
1:1/4:4	6.75	67	1.00
1:1/2:2	12.40	250	0.92
1:1/2:3	8.80	175	0.98
1:1/2:4	6.75	135	1.00
1:1/2:5	5.40	110	1.00
1:1:3	8.60	345	0.95
1:1:4	6.60	270	0.98
1:1:5	5.40	210	1.00
1:1:6	4.50	180	1.00
1:1 1/2:3	8.10	485	0.90
1:1 1/2:4	6.35	380	0.94
1:1 1/2:5	5.30	320	0.98
1:1 1/2:6	4.50	270	1.00
1:1 1/2:7	3.85	230	1.00
1:1 1/2:8	3.40	205	1.00
1:2:4	6.10	490	0.90
1:2:5	5.10	410	0.94
1:2:6	4.40	350	0.98
1:2:7	3.85	310	1.00
1:2:8	3.40	270	1.00
1:2:9	3.00	240	1.00

Fig. 5-1. Quantities of materials needed to make 1 cubic yard of various mortar mix ratios.

included in these premixes, and you only have to add water. Numerous brand names are available, and there are various sizes of sacks. Sakrete, one of the best known manufacturers of premixes, produces a mix which results in a mortar with a compression strength of about 1250 pounds per square inch, an approximation of a Type N mortar. The amount of work that you can accomplish with a sack of mortar depends upon both the sack size and the thickness of the joints you plan to use. For example, Sakrete's 80-pound sacks of mortar mix will lay 65 bricks or 27 standard concrete blocks (8 × 8 × 16) if you use 3/8-inch joints and don't splatter too much mortar onto the ground.

The Proper Consistency

The mortar mix you finally choose is always a compromise. The less lime there is in the mix, the stronger the mortar will be; but on the other hand, without the lime, the mortar is unworkable. If there is a great deal of sand in the mix, the mortar will be more difficult to work and will have less ultimate strength. To a point, the less sand used the more rugged the mortar will be. But the cost rises as more and more cement is needed. The smaller the amount of water used for a given batch, the greater will be the strength of the mortar, but you have to have sufficient water to make a plastic, workable mix.

By far the commonest reason for mortar mixes which don't behave they way they are supposed to is improper measuring of the dry ingredients. Frequently the difficulty is too much sand, making a given type of mix too harsh and grainy to be easily worked. There is no way to mismeasure the ingredients that are purchased already bagged, but errors do occur when the sand is blended by the shovelful, without precise measurement. Inaccurate quantities of cement and lime can also cause problems. If there is too much of either or both, the mortar turns out to be like cold oatmeal and sticks firmly to the tools. This is called *fat* mortar. *Lean* mortar is the opposite, it doesn't have enough cement and/or lime. Lean mortar, which doesn't stick to anything, is very difficult to work.

Assuming that the mortar ingredients have been accurately measured by using a bottomless box or some other container of known quantity, then the next variable is water. Too little water means an unworkable mix, too much water likewise means an unworkable mix. But it's different than mixing concrete; there are no recommended amounts of water best for a given type of mix, so you can't mix "by the numbers" or "by the book." Mortar has to be mixed entirely by sight and feel, until the proper degree of workability and plasticity is reached for the job at hand. The procedure for attaining this custom-mix depends upon several factors.

The preferences of the mason play a large part, and so do weather conditions. The more moisture there is present in the sand, the less water will have to be added. The absorption qualities of the masonry units has a bearing. And of course the workability must be just right—if there is too much water in the mix, the mortar will zip right off the trowel, or slither off the ends of the units while they're being buttered. On the other hand, if the mix is too dry it will be crumbly and won't have proper adherence characteristics.

Prescribing a proper mortar mix is a virtual impossibility, since so much depends upon a sense of rightness and a correct "feel," but we can try. The consistency of a good mortar mix is smooth and even, with no lumps, and is much like a thick mud. In fact, "mud" is a widely used slang expression meaning mortar. The mix must be plastic but stiff—like a thick cake frosting—so that you can shape it and have it remain in place. It must not compress or squish out of the joints when several courses of masonry units are laid atop it. On the other hand, it must settle out and reform its surfaces under that weight just enough to establish a solid bond between the masonry units. It must also be plastic enough to squeeze to the face of the joint as the mason pushes down on the masonry unit to achieve the proper joint thickness. The mix must be sticky enough to remain on the trowel for a short while, yet won't cling like a glue when the mason gently urges it off. After the mortar is flicked off the trowel, the trowel surface should be free of any residue. Still, the mortar must be pasty enough to adhere to the vertical edges of the masonry units as they are set in place. Furthermore, mortar should also spread easily and smoothly.

How do you know when your mortar mix has reached this stage? Experience is really the only answer. After some trial and error, you will begin to get the feel for a good mortar mix and you will be able to tell if the mortar is too fat or too lean, or has some other problem. One good way to find out what it's all about is to strike up an acquaintance with an experienced mason who can show you what a good batch looks like and let you get familiar with the nature of it. Watching a mason work for a while is also an excellent way to learn the rudiments of masonry unit construction.

Mixing the Ingredients

Mortar can be mixed either by hand or in a power mixer. The latter method obviously saves time and labor, not to mention effort.

By and large, power mixing will also produce a better mix, because the more thoroughly worked mortar is more plastic than that which can be easily obtained by hand mixing. Up to a point, the more you mix, regardless of the method, the more plastic and workable the mortar will become. The tendency is to stop too early when mixing by hand, because it's a lot of labor. If you use a power mixer, the batch should be mixed for a minimum of 3 minutes, and preferably a bit more. Mixing by hand should go on for a longer time, at least 5 or 6 minutes.

To mix a batch of mortar, the first step is to apportion the materials. Generally you can figure that the amount of mortar obtained from a given type of mix will be roughly equal to the volume of sand used. With a power mixer, the usual procedure is to drop in the cement, then the lime, and then the sand. This should be done while the mixer is running, and the dry materials allowed to turn for a bit. Water is added and the mortar given time to mix thoroughly. If you are using a tub, hoe box or mixing platform, dump in the cement, lime and sand and mix them all thoroughly with a hoe or shovel. Draw the dry materials into a pile, make a hole in the center, pour in your water and mix again. Continue this process until the mortar has reached the right consistency. With either method, the water should be added a little bit at a time, until you have the proper mix. There is a danger of adding too much water, because as the mix approaches its final stages even a small amount of water can make it soupier than you want. Trying to add more dry ingredients to a batch of mortar that is too wet, to thicken it up, does not work out well at all.

Mortar cures in the same manner as concrete, by hydration. It too will harden or set up somewhat as the batch lies waiting to be used. This hardening process, which is only temporary, may begin almost immediately after mixing if the weather is hot, windy and dry, or may not occur to any appreciable degree at all in cloudy, humid, still and cool weather—that is, up until the time the mortar begins to cure. You can usually assume that curing by hydration will begin 2 1/2 hours after the batch is mixed. This means that you should not prepare any more mortar than you can use in that length of time.

In the meantime, if the mortar starts to harden up and becomes difficult to work, stiffens and loses its plasticity, you can do a bit of remixing. This process is called *retempering*, and should be done to

Fig. 5-2. Retempering mortar on mortarboard by adding a trickle of water and remixing.

mortar as soon as any stiffness sets in. It may be done either on the mortar board or while the mortar is waiting in the mixing container. Dribble a small amount of water over the batch and mix thoroughly (Fig. 5-2). If the mix still seems stiff, add a bit more water, but be careful to retain the proper consistency and not to turn the mix to soup. Remember the time limit of 2 1/2 hours, which can sometimes be stretched to 3 hours under cool conditions. When in doubt, throw the mix out; work up a new batch and start over.

Additives

Most of the time nothing is added to the principal ingredients of the mortar mix, but there are occasions when some additives prove of value. One such additive is a waterproofing compound, mixed into mortars which will see service in structures exposed to heavy weather or amounts of moisture which are well above average. These compounds serve to make the finished mortar denser than normal so that moisture absorption is inhibited. This helps to preserve the integrity of the structure, especially where hollow masonry units are used in the assembly.

A new additive has recently been put on the market to give extra strength to mortar, since mortar joints constitute the weakest points in a masonry unit structure. This additive is epoxy, a chemical compound which imparts great strength and bonding power to the mortar, just as it does to glues and adhesives. In fact, some epoxy

176

mortars are stronger than the masonry units themselves; thus the problem is neatly reversed. Adding epoxy is usually not necessary and should only be done where the added strength is important and the added cost is not. There are various brands and different mixing methods, so if you feel you have a need for an epoxy additive, get specific information and recommendations from your building materials supplier.

Perhaps the most commonly used additives are coloring agents. These are mixed into the mortar to provide joints which either blend in completely with the masonry units, or contrast with them and give a decorative effect (Fig. 5-3). These additives are mineral oxide pigments (dyes and other coloring agents are never used) in combination with white portland cement, and (preferably) with white or very light-colored sand. A wide range of earth tones can be obtained, such as true gray, blue-gray, yellow, off-white or cream, pink, and red.

Mixing for contrast is simple enough, and involves only mixing up a small trial batch and adding pigment until the desired tone or shade is reached. Make a note of the exact proportions, and mix all subsequent batches in exactly the same way. Mixing to match is often a much more difficult process, as it is with paints. This may require a series of trial batches, perhaps using a combination of pigments to arrive at the best-matching color. Sometimes the wet mortar batch will have a somewhat different tone than the cured mortar, so a number of experiments must be made.

Mineral oxide pigments are added to the mortar at the dry-mix stage, and thoroughly blended with the other ingredients until the coloration is perfectly consistent throughout. Even so, some streaking will probably show up as the water is added. Thorough mixing for several minutes will eliminate the condition.

Spreading Mortar

Trying to explain verbally and without any visual demonstration exactly how mortar should be spread is almost as difficult as trying to describe a proper mortar mix. If you could watch a mason in action for a short time, you would quickly see how the job is done, although a certain amount of practice is needed to gain confidence, develop dexterity, and become skilled. More details about mortar spreading

Desired Color	Mineral Oxide
Natural gray	-----
White	White portland and sand
Dark gray Blue-gray Black	Germantown lamp-black, black oxide of manganese, or carbon black
Brown	Burnt umber, or brown iron oxide
Brown-red Brick red Pink Salmon	Red iron oxide
Red sandstone Purple-red	Indian red
Bright red Vermillion Carmine	Mineral turkey red
Buff Cream Yellow	Yellow ochre or yellow iron oxide
Blue	Ultramarine blue
Green	Chromium oxide

Fig. 5-3. Mineral pigments used to mix various colors of concrete, mortar or stucco.

will be discussed throughout the text as the occasion arises, but basically, this is the procedure.

To begin with, the mortar board with its small pile of freshly mixed mortar—a bucketful or so—should be placed as close as possible to the work area and positioned so as to necessitate the fewest number of motions and the least amount of effort in getting the mortar from the board to the masonry units being buttered. The mortar board should be set at a convenient height to minimize stooping or reaching. All of this is particularly important to the mason who has a long day's work ahead of him and wants to pace his production properly and reduce fatigue as much as he can.

The mortar is applied with a mason's trowel or a buttering trowel, or both, depending upon the preferences of the worker and the size of the masonry units. The first step is to remix the mortar on the board with the trowel, to keep the mix plastic and dispel any slight surface hardening. Most masons do this automatically every time they go back to the board for more mortar. Then the mortar is scooped into a pile and loaded on the trowel. A quick flick of the wrist with the trowel held level flips the excess mortar back onto the pile and evens the load on the trowel for easy placement (Fig. 5-4). The amount of mortar left on the trowel after the wrist snap should be equal to the amount of mortar needed for laying on a certain section.

Fig. 5-4. A practiced flip of the wrist leaves just the desired amount of mortar on the trowel.

The ability to accomplish this is something that you can only develop by experience.

The next step is *throwing the mortar*, and consists of a series of quick wrist and forearm motions. The trowel, held horizontally, starts at the farthest point to be mortared. With a series of flips the mortar is unloaded onto the masonry surface in a long string of remarkably even dimensions. By the time all of the mortar is unloaded, the trowel is upside down.

The next motion brings the trowel back to the starting point, still upside down and ready for the next step, which is called *spreading*. This consists of drawing the trowel point back down through the line of mortar, creating a furrow in the center and pushing the edges of the mortar outward slightly. If the job is done right and the correct amount of mortar has been put on, the layer (called a *bed*) will be quite uniformly shaped and no mortar will be hanging over the edges of the masonry units.

If there is overhanging mortar, then the next step is called *cutting off*. This involves running the edge of the trowel blade, held vertically and at a slight compound angle to both the masonry unit and the mortar line, along the edge of the masonry, cleanly striking off any excess mortar.

The above steps are most often used with solid and rather narrow masonry units, or in double lines of mortar bedding on larger solid units or on those which have irregular faces. Though the double-line method can also be used with hollow-core masonry units, such as concrete block, there is a second method. In this case, the trowel-load of mortar is poised above the block at the farthest of the corners to be mortared. Then the trowel is quickly turned sideways and lowered before the mortar can slip off the blade, contacted against the edge of the block, and scraped down and back at the same time along the face edge. If done properly, this leaves a ridge of mortar standing along the edge for the full length of the block. The other edge of the block is treated in the same manner. A dexterous mason can do one edge with a forehand sweep and the other with a single backhand movement.

Mortaring only the edges is called *face shell bedding*, or *face bedding* (Fig. 5-5A). If the webs of the unit (the cross sections that tie the block faces together), or, in the case of solid units, the entire

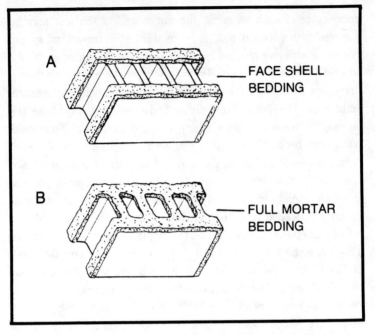

Fig. 5-5. The two methods of applying mortar to concrete block.

surface, is to be mortared, small gobs of mortar of just the right length and thickness are flipped onto the webs or the full surface with a snap of the wrist. This type of mortaring is called *full mortar bedding* or just *full bedding* (Fig. 5-5B). In either case, once the mortar is laid in place, it need not be touched again, providing that the right technique was used. If not, then a certain amount of cutting off with the trowel blade or spreading with the trowel tip must be done. But the mortar should not be fussed with any more than is absolutely necessary.

All of the previous explanation referred to the mortaring of masonry units which had already been integrated into the structure. The bedding may be placed only on one masonry unit at a time, but more often is spread along a whole series, composed of as many as a dozen bricks or half a dozen blocks. The procedure which will be described next, however, is performed upon each masonry unit prior to setting it into the structure.

The first maneuver is to *butter* one end of the unit to provide the vertical mortar joint which is called a *head joint*. With smaller masonry units such as brick, the entire end of the unit may be buttered with a layer of mortar scraped from the trowel and spread into place with an even sweep. Larger units with flanged ends, like concrete blocks, are buttered with a line of mortar along each flange. Large square-ended units are buttered with a line along each vertical end corner. The mortar is usually applied in the same way as the face bedding, with a downward scraping sweep of the trowel. Small units which can be handled singly, like small brick, are usually end-buttered one at a time as they are set in place. Larger units like concrete blocks, which require both hands to maneuver, can be stood up on end and buttered half a dozen at a time, or in whatever quantity is needed to cover the course bed that was just laid.

After end-buttering, the unit is emplaced as soon as possible. This is done by placing the unit in its proper position along the length of the course and then aligning it vertically with the previous courses. It is carefully set onto the fresh mortar bed slightly ahead of its ultimate resting place. The unit is at the same time pushed gently up against its neighbor, spreading the mortar on the buttered end into a full joint. Then the unit is gingerly tapped into final place with the trowel handle, checked for vertical and horizontal alignment. Last of all, the joint thickness is adjusted.

While this is going on, a certain amount of mortar will squeeze out of the joint all around the unit. This is exactly what should happen, and means that you have a fully-filled mortar joint. If no mortar squeezes out and there are obvious gaps in the joint that you can see into, you have two choices. You can stuff some mortar in the spaces and hope for the best, or you can remove the unit, scrape the mortar away, and start over again. The latter course is the more frustrating, but is the better one.

After the unit is in its final lie, your next step is to *strike* the excess mortar that has squeezed out of the joint. This is done with the side of the trowel, using a combined forward and slightly upward scooping/scraping motion which leaves the extra mortar on the trowel. The scraps can be chucked back onto the mortar board and mixed in for reuse.

The above steps are basic and applicable to practically all kinds of masonry unit wall-type construction, except perhaps when fieldstone is involved. With fieldstone the steps are not quite so precise, and making the joints is a matter of throwing and buttering as needed and filling voids where necessary to make up for the irregularities of the stone. These procedures are carried on throughout the project, except at two principal points.

The first point is at the starting course, the line of masonry units laid directly atop a footing. Here a full, wide and thick bed of mortar is thrown down, spread out, and furrowed with the trowel point to make a broad base. The units are laid in the mortar bed one by one with their ends buttered in the usual fashion. They are carefully positioned, leveled and, if necessary, graded. They must also be brought perfectly plumb. The excess mortar that is squeezed out of the bottom joint is left in place if the joint will be below grade, or cut away if the joint will be visible.

The second point occurs at the stage where closure units are installed. A closure unit is the last unit to be laid in each course; it closes up and completes the course. The only difference between this unit and any of the others lies in the way it is treated: all of the edges of the last opening and all of the edges of the closure unit are buttered with mortar. When the unit is slipped down into place, there is sufficient mortar along all vertical edges to make a good joint even if some of the mortar is knocked away. If the mortar mix is of the right consistency, the buttered joints will squish and slip together without crumbling or dropping off.

GROUT

Grout is a term which is used in several contexts in the building trades. By definition, grout is a thin mixture of cementitious materials and water which solidifies soon after emplacement in the joints and voids between solid units, and which serves to bond the units into a whole. As used in masonry unit construction, grout is a thin mortar for pouring into the voids and cores of walls and similar wall structures. It can be used by itself to give further strength to the structure by binding the masonry units together firmly. It is also frequently used to embed steel reinforcing rods and mesh inside a structure.

In another use, grout is the material which is worked into the spaces between various kinds of tiles and pavers. Until recently the most commonly used grout for ceramic wall and countertop tiles, and sometimes floor tiles as well, was a powdery blend of white portland cement and an extremely fine white sand; it strongly resembles a baker's flour. A similar type of grout made from standard gray portland cement is sometimes used on floor tiles and pavers. Because of the failure of this kind of grout in many situations, new and more effective materials have come to the fore. There are two main types; furan resin, and epoxy. Though the ingredients are different, the principle and the applications remain the same.

To further confuse the issue, the term grout is also sometimes used to indicate the final coat of plaster applied to interior walls and ceilings. And there are two other uses of the term grout, though neither is really correct. When pavers, flagstones, and some types of tile are set as flooring, they are often placed upon a bed of ordinary mortar. Though this mortar is no different than that used for masonry unit construction, it is sometimes called grout. Furthermore, when a concrete mix which is used for casting small articles and thin sections in specially-made casting forms is made rather thinner than normal for ease of pouring, it is sometimes called grout.

Filling grout is made the same way as ordinary mortar, and includes the same ingredients, but with a great deal more water. There are two different classes of grout used for filling purposes: fine, and coarse. The only aggregate in fine grout is sand, and this mix is used to fill voids smaller than 2 inches in size. A coarser aggregate such as *pea gravel* is used along with the sand in coarse grout. This increases the workability of the material and decreases the cost. Joints and voids up to 4 inches can be filled with this material. Larger voids and spaces are best filled with an appropriate type of concrete mix. Grout is akin to mortar in two respects: the ingredients must be accurately measured and the mix must be applied within 2 1/2 hours.

As with other cementitious materials, the trick to mixing grout is deciding on the correct amount of water to gain the right consistency. Grout must be liquid enough to pour readily from a bucket (it is often pumped into place where large quantities are used), but not be so wet that it won't mix or cure properly. The easiest way to determine whether the mix is correct is to make a slump test, the same as for concrete mixes. When the slump cone is removed, the pile of mix should slump anywhere from 8 to 11 inches. The 11-inch slump is quite liquid and will pour readily, while the 8-inch slump will flow less freely. After the mix is poured into the voids, it should be rodded and churned a bit to make sure that any air bubbles are dispelled and that the grout is worked into all the chinks and fissures.

Tile grout is handled somewhat differently. Gray cement grout can be made like an ordinary mortar, using fine sand, but extra water. The mix is usually much stiffer than a pouring-grout mix, but less so than most mortars. This type of grout would be used primarily for flooring applications and for a few wall tile applications as well. The white tile grout is mixed in much the same manner, but uses an extremely fine white sand. Ceramic countertop and wall tile is most often filled with one of the commercial premixed and packaged tile grouts, many of which contain additives of one sort or another. Only water need be added, and complete instructions are included on the package. The resulting mix actually looks, feels, and acts more like plaster than mortar.

Although filling grout is dumped into place with a bucket or a scoop, tile grout is not. There are various methods of application and sometimes a special curing process to follow as well. But by and large the procedure is one of depositing some grout and spreading it into the cracks with a trowel or squeegee. While the grout is setting up, as much excess as possible is carefully brushed away. After a certain curing time has passed, and depending upon the nature of the tile, the residue is either sponged away with water or is cleaned with an acid treatment.

STUCCO

Stucco is not a structural material, but rather a type of finish. Though often chosen because of its long life, remarkable weather-resistant qualities and low maintenance requirements, it is also used because of its decorative and architectural attributes. Stucco does not stand alone; it is applied as a coating over a structural base or framework which can be fabricated from many kinds of building materials. Stucco can also be used to refinish an existing surface. It makes little difference whether that surface is in good condition or poor, so long as it is structurally sound.

Stucco is sometimes referred to as a plaster, and in the sense that it is a pasty composition used to cover an underlying surface, this is accurate. Also, in the days of excessive architectural ornamentation, the elaborate plaster moldings and adornments crafted on walls and ceilings were called stuccowork. Today, however, the meanings have shifted. Plasterwork is the covering applied to interior walls and ceilings, while stuccowork is exterior wall covering. A plasterer applies the former, a stucco mason the latter, though either job could be done by either artisan.

Stucco Mixes

Stucco is essentially a mortar, and uses the same ingredients. Its foundation is the standard portland Type I cement, though Type II can be used also. Most stucco jobs are done with three coats of material; the *base* or *scratch coat,* the second or *brown coat* and the *finish coat*. In a few instances, particularly if the job is quite small, only two coats are used: an extra-thick base, and the finish coat. The standard gray portland cement can be used for the scratch and brown (which obviously isn't) coats. If the final finish is to be a neutral gray, then gray portland can also be used for the finish coat. More often than not, though, a white portland is used for the finish coat, either as is or with the addition of mineral pigments for coloring.

Standard masonry cement can also be used in stucco mixes. In some areas of the country, notably those places where stucco is in widespread use, another type of cement is also available. This is called *stucco cement*, or *plastic cement*, and consists of an exceptionally finely ground cement designed for this purpose.

Stucco must work easily, cling well, and retain its shape and position after application. This means that a plasticizer must be added to the mix, and the most common choice is hydrated lime. The working rule is that an amount of lime up to 10 percent of the weight of the portland cement can be added to the mix, but no more. Since in most cases a high degree of plasticity is desirable, the maximum amount of lime is usually added. If masonry cement is chosen instead of portland, however, no lime is added, because the lime is already included in the masonry cement. Nor is lime added to stucco or plastic cement; the mix will be sufficiently plastic as is.

The greatest difference between mortar and stucco lies in the kind of sand used. While mortar sand particles may run as large as 1/4 inch, the maximum allowable size for stucco is 1/8 inch.

Another critical point is that stucco sand must be evenly graded through the entire range of particle sizes from 1/8 inch on down. If the sand is not well graded, there will be a larger than necessary number of tiny voids between the grains. This means that additional water and cement must be used to fill those voids. The extra cement means extra cost, to be sure, but that's not the central problem. As with concrete and mortar, the more water you use in a given mix, the weaker the finished product. With stucco, maximum strength is of great importance. Stucco is a thin shell subject to all manner of stresses and strains from vibration, expansion, contraction, and weathering. So it doesn't take long for a weak stucco to crack or craze. Well-graded sand, then, demands less water, which means greater strength and no cracks from inherent weakness.

As with any cementitious mix, the water used must be clean. Nearly any potable water is satisfactory for mixing stucco, unless there happens to be a high sulfate content. Acids and alkalies are detrimental, and salt water should not be used.

There are two more or less standard mixes for stucco, one for the scratch and the brown coats, and another for the finish coat. For the scratch and brown coats, mix 1 part cement, 3 parts sand, and 1/10 part hydrated lime. If you are using masonry cement instead of portland, keep the same proportions of cement and sand, but omit the lime. For the finish coat, the proportions are 1 part cement, 2 parts sand, and 1/10 part hydrated lime. Again, if you are using masonry cement, omit the lime. Also, unless the finish color is to be gray, use a white portland instead of a gray portland.

Though these mixes are standard, certain variations may be introduced, depending on the circumstances. For instance, you don't have to use the maximum 1/10 part lime; you can cut back on this ingredient if you wish. If you do cut back, however, the mix will not be as plastic or workable. The amount of sand may also vary. The proportion stated is predicated upon the use of damp, loose sand of a broad and uniform gradation. However, the moisture content, gradation range, and unit volume of sand can vary according to locality. Though 3 parts is recommended, as much as 5 parts may be necessary in some cases. The danger is in using too much sand rather than too little, and a few small trial batches should be mixed to help you make your selection. Once you determine the suitable proportions, make a note and stick with those figures until you change batches of sand.

Mixing colored stucco requires considerable care and close control of the materials. Best results are obtained by using well-washed and well-graded white stucco sand, along with the white portland cement. The greatest problem, however, lies in the coloring process itself. As with mortars and concretes, the coloring is done by adding certain mineral oxide pigments. Streaking and mottling can only be avoided by rigorous mixing at both the dry and the wet stages. The biggest difficulties arise in color matching on large projects, such as walls. The stucco is usually mixed in a long series of small batches, even though the job may be relatively large in terms of square feet, and trying to create an identical color tone in each batch can be next to impossible.

When mixing from scratch, all of the materials have to be of uniform coloration, and preferably the entire quantity of each material should have come from the same bank, or processing-run. The proportions of ingredients must be held constant from batch to batch, particularly in the case of the pigments. Only small quantities of pigment are introduced, up to a maximum of 9 pounds of pigment per sack (one cubic foot) of cement, but frequently a good deal less. The pigment should be measured out not by volume, but by weight. Once the desired color tone is attained by means of trial batches, a note should be made of the formula and the exact same weights should be used in subsequent batches.

As with other kinds of masonry materials, there is a way to avoid all the fuss and bother of mixing from scratch. You can use what is known as *stucco finish*, a premixed package of ingredients to which you need only add water. Specific instructions for mixing are included on each package or sack. Using a premix is particularly helpful if you want a colored stucco. Provided that the supplier has the shade you want, you can avoid all the problems of matching colors between batches, because premix batches will always come out the same. The manufacturer makes his product to strict standards and his quality control is superior to anything that can be accomplished in the field where the conditions are mostly uncontrolled and only basic equipment is available. And though the cost of materials may be somewhat higher when using premixed stucco finish, the overall amount of labor may well be less, and the final product better. There are times, too, when making up the stucco base coats from scratch and utilizing the premix for the finish coat is a good policy. Much depends upon the conditions of the job, availability of various materials, costs, and other factors.

Mixing Stucco

Even for a large-scale project like the exterior of a residence, stucco is usually mixed in a succession of small batches and applied continuously, working from section to section of the project. Mixing can be done by hand in a wheelbarrow, tub, hoe box, or on a mixing platform. Any of these works well for small projects, such as covering an entryway arch or a garden wall. If stucco mix requirements for the project run to much more than a couple of cubic yards, however, hand mixing will involve a tremendous amount of toil. Don't forget that once the stuff is mixed, it has to be applied, and that's no easy chore either. As with mortar, you can figure that the yield of each batch will be roughly equivalent to the volume of the sand. This means that if you mix in a wheelbarrow, you'll need probably seven or eight batches per cubic yard.

A power mixer will make job far easier, and though many professionals use mixers which are of a design slightly different than the standard small cement mixer, any type of power mixer can accomplish the task. If you don't have one, you may be able to borrow or rent one. Make sure that the mixer drum is thoroughly clean and won't deposit chunks of dried mortar or concrete into your fresh stucco mix.

The mixing process starts with the dry ingredients: first the cement; then the lime, if it is to be used; and then the sand. Measure everything carefully. If you are coloring a white portland for a finish mix, the white portland and the white sand go in first, followed by the required amount of pigment. The ingredients should be mixed for several minutes, so that the pigment disperses evenly throughout the material and coats all of the aggregate particles. After the dry ingredients are well mixed, the water is added and mixing continues for at least another 5 minutes. If the mixing is done by hand, the period should be longer. Incidentally, power mixing gives a more uniform and consistent mix, especially when pigments are an ingredient.

Running a series of trial mixes is almost essential when working with stucco. These batches can be small, hand-mixed ones, just large enough to be able to give you a good idea of exact quantities needed. A number of batches may be necessary to develop the proper shade for colored finish coats. But whether the stucco is base coat or finish, several important characteristics of any given batch should be noted.

The stucco must be plastic and thoroughly workable, so that when you put it in place it will stay there, and, furthermore, will retain whatever shape you give it. The mix must be sticky enough to adhere to the base surface, but not so sticky that it clings to the tools. Once in place, the mix should not sag, but remain perfectly true. Nor should it "leak"—exude moisture which will form in small pockets or actually seep out of the material. Once you arrive at a proper combination—neither too much nor too little of either sand or water—make a note of the exact proportions used and mix subsequent batches in the same way. If it takes a couple of days or longer to complete the project, or if you receive a new batch of sand, it's a good idea to run more trial batches, just to double-check yourself. This is especially true if the stucco being applied should suddenly begin to handle poorly.

Base Preparation

A stucco finish can be applied to just about any kind of base, with one exception. Stucco cannot be put on over a base surface of plaster, or of any material containing gypsum, because a chemical reaction takes place between the gypsum and the ingredients of the stucco which causes eventual failure of the bond between the two layers. Sooner or later the stucco will crack and peel away. A base of any nongypsum material, however, is all right, provided that it is in good repair and has been properly prepared.

Stucco is often applied over wood frame structures, either new or old. The exterior walls of a house are the most common example, but the same principles of preparation and treatment are applicable to other structures, such as an arched entryway to a formal garden or a free-standing wall. The wood skeleton is first covered with a sheathing such as plywood or sheets of impregnated building board. Then a layer of roofing felt or tar paper is nailed or stapled to the sheathing. Lightweight felt (15-pound or 30-pound) is sufficient, and should be laid in horizontal strips with a narrow overlap between strips. This provides a waterproof barrier between the sheathing and the stucco.

Stucco will not adhere to a smooth surface, so the next step is to provide something to which it can cling. The material used is *wire lathing*, and it is available in a number of designs. One of the most common types is *stucco netting*, which looks much like ordinary chicken wire, and is sold by the roll. Lengths of the proper size are cut from the roll and fastened in place in much the same manner as the roofing felt was. The wire is attached to the surface of the structure with *furring nails*, double-headed nails manufactured so that when they are driven fully home, the stucco netting will be held about 1/4 inch away from the structure surface. This allows the base coat of the stucco to completely surround the lathing, creating a rigid covering that cannot part from the structure.

There are also several types of metal lathing known as *expanded metal*. These are made from flat sheets of metal stock which are punched or slit by machine and then pulled apart into a diamond-shaped mesh. A variety of mesh sizes and shapes is available, in flat sheets of assorted dimensions. Like stucco netting, expanded metal is cut and fitted into place, and then attached with furring nails. This type of lathing works best on flat surfaces, while the stucco netting can be easily bent to conform to the contour of virtually any surface.

Stucco can also be applied to all sorts of masonry walls, but the job is easiest if the masonry structure is built especially to receive a stucco coating. If the proper steps are taken, metal lathing is not necessary. Most concrete block walls are rough-textured enough to accept stucco, and some poured concrete surfaces may also be satisfactory. The surface should be thoroughly cleaned so that there is no dust or other foreign material present. After the surface has dried, spray a fine mist of water on the structure and note the results. If the water immediately beads and runs off, the surface is not porous enough to accept stucco. If the water soaks in, stucco will adhere. On the other hand, if the surface appears to absorb great quantities of water, the structure may absorb too much moisture from the fresh stucco mix during application, which would result in a loss of strength in the stucco and a weakened bond. The remedy in that case is to lightly spray the surface immediately prior to stucco application, but refrain from a lengthy soaking.

A better bet with poured concrete is to provide mechanical keying for the stucco when the structure is built. This can be done during the form-building stage by tacking strips of molding to the inside surfaces of the forms. When the forms are stripped, the concrete surface will have a series of grooves into which the stucco can key. Unformed concrete surfaces can be roughened up during the finishing stage with the tines of a steel rake or some similar implement. The furrows and scratches will serve as keys. Cured concrete surfaces can be roughened up with a hammer and cold chisel; grooves and pits can be produced by chipping out the surface.

When concrete block or brick structures are built with a stucco finish in mind, the joints between the masonry units can become the keys for the stucco layer. As the structure is assembled, the joints should be raked out with a raking tool, freeing them from mortar to a depth of 3/8 inch or even 1/2 inch. This will not reduce the strength of the masonry unit structure to any appreciable degree, and will provide plenty of keying for stucco.

Where good keying does not exist, or where there is some question as to whether the keying or the porosity of a given surface will be sufficient to hold the stucco coating, there is another method you can use. Cover the surface with a bonding agent which will adhere tightly to the structure surface and form an interface between the structure and the stucco. The bonding agent, of course, must be able to bind the stucco firmly. There are special bonding-agent mixes formulated for this purpose, and you can purchase them at masonry supply outlets. You could also apply a *dash bond*. This consists of a paste of 1 part cement and 1 part fine sand mixed to a fairly loose consistency, but not sloppy. Its flow characteristics should be about the same as those of a thick texture-paint. The dash bond should be liberally applied to the entire structure surface with a stiff brush, leaving plenty of brush marks, grooves, ridges and similar surface irregularities. Allow the dash bond to cure thoroughly before applying the stucco finish.

There are plenty of surfaces on which the foregoing methods will not work. Old brick is often suspect because of an unstable surface. Ceramic tile or glazed stonework is too smooth. Metal is completely nonporous, and there is no effective way to install keys. In circumstances of this sort, the recourse is to go back to the metal lathing system. Irrespective of the composition of the structure surface, the first step is to apply a layer of roofing felt. Then the metal lath is fastened in place. The kind of fasteners used depends upon the structural material, but some type of furring nail should be used to attach the mesh or lath, holding it away from the roofing felt by 1/4 inch to 3/8 inch. Then the stucco is applied in the usual fashion.

When a stucco covering is to be applied to an existing surface, first check the structure itself to make sure that it is perfectly sound. If there is an existing exterior covering, such as the siding on a house or some other material which can be removed without too much effort, then this covering should be stripped away to the sheathing (or other subsurface) and the wire lathing attached in its stead. Repairs should be made to the sheathing or to the structural framework as necessary, to provide a rigidly solid base.

Whether the structure is new or old, the lathing should cover every bit of the surface which will be stuccoed, and be solidly anchored at all points. Corners and other angles should be beaded with wood or metal strips made for the purpose. Remember that the stucco must form an integral part of the structure and be applied so that no moisture or weather can seep into cracks or behind the coating. You should arrange the lathing and beading with this in mind. If moisture gets in anywhere, the inevitable result is cracking and eventual deterioration.

Applying Stucco

Putting on a stucco finish is not an especially difficult task, though a considerable amount of physical effort is required. The technique of emplacing the fresh mix (called *throwing*) does take a little time to learn, but with a small amount of practice you will quickly get the hang of it. You can experiment a bit during the process of applying the base coats, and if you make a few minor errors, little harm is done so long as you get the right thickness and a good bond. The base coat will be covered with the finish coat anyway, so slight imperfections will be hidden.

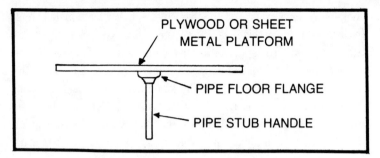

Fig. 5-6. Do-it-yourselfer's plasterer's hawk.

Another point worth noting is that stucco is seldom applied in a perfectly smooth, wave-free surface such as would be done with interior walls or ceilings. Although the stucco maintains a relatively uniform thickness throughout, its surface is normally ripply and wavy and well-covered with tool marks. Frequently various decorative effects are worked into the surface at the finish stage, such as stippling, swirling, or any one of a hundred other simply-executed designs which leave a roughened or textured surface. This means that the relatively unskilled and inexperienced do-it-yourselfer can expect to make a stucco application without any undue difficulties and with results comparable to those of a professional. The amount of expended effort and elapsed time may well be greater than for a professional job, but this usually is not of prime concern to the home mechanic.

The tools you need for applying stucco are few. A mortar board is handy to dump the fresh mix onto and to work from as you transfer the fresh mix to the working surface. A mason's metal float about 4 to 5 inches wide and 10 to 12 inches long is ideal for emplacing the stucco mix. You might also like to use a *plasterer's hawk*, which you can easily make yourself (Fig. 5-6). The platform portion is a scrap chunk of plywood with a good upper surface, about 12 to 16 inches or so square. The stubby handle should be solidly attached and can be made from a piece of 1 1/2-inch wood round or a short length of metal pipe threaded into a floor flange and attached to the underside of the platform.

To use the hawk, transfer a heap of fresh mix from the mortar board onto the hawk and trowel the mix from the hawk onto the stucco surface. Though there are more motions involved, this intermediate step can often make the job easier, depending upon the circumstances.

The scratch coat is the first one to be applied, and should be spread and compressed into place at the same time. Use plenty of pressure so that the mix completely surrounds all the strands of the metal lathing and fully fills the space between the lathing and the roofing felt, leaving no voids or airspaces. Though the gap between the metal lathing and the roofing felt is only about 3/8 inch, the thickness of the scratch coat should be approximately 1/2 inch. If the lathing is made from expanded metal, most of the mesh will be completely covered. With stucco netting, which is more flexible, more of the wire is likely to show through. Even so, for a good job 80 to 90 percent of the strands should be well embedded in the stucco.

As soon as a section of the scratch coat has been emplaced, the surface should be scored or scratched with a *scarifier*. This leaves a continuous series of horizontal grooves in the coating to which the next coat will key. You can purchase a scarifier (also called a *scratcher*) at a masonry supply house, or you can make your own. Chop up a few pieces of fairly stiff steel wire or some old coat hangers into equal lengths. Eight or ten pieces will do nicely. Collect them into a bundle and wrap one end tightly with electrician's tape to form a handle. Then bend the individual lengths to make a flat fan shape, with all of the wire tips in the same plane. Score the stucco by drawing the scratcher across the fresh surface in long lines. Use enough pressure so that the grooves are fairly deep and well formed. You should be able to accomplish this without dislodging any material.

The second or brown coat should be applied soon after the scratch coat has been completed. The interval between coats varies somewhat according to the type of base and to the weather conditions. In the case of a solid backing with no metal lathing, you may be able to start applying the brown coat after as little as 5 hours. Where a roofing felt backing and metal lathing is used, the time will probably extend to about 2 days. Either way, begin when the scratch coat has entered well into the curing stage, but is not yet completely hard. The right condition is known as a *firm set,* and is difficult to describe.

When the interval between coats extends to about 8 hours or more, the brown coat should be sprayed briefly with a fine mist of water from a garden hose. The surface should never be drenched—only lightly dampened. If water drains from the surface, you are overdoing it. The wetting operation can be done in segments, while the brown coat is being applied.

Although the scratch coat can be put on piecemeal and without too much regard for the appearance of the surface, greater care should be taken with the brown coat. The application of the brown coat should be done either all at once, or in complete sections, so that there is no visual evidence of the work interruption. This will give you a smoother and a stronger job.

The process for applying the brown coat is much the same as for the scratch coat. Instead of using a metal trowel, however, the mix is applied with a wood float, and no great amount of pressure is needed. The mix should be spread with even strokes and floated smooth as you go along. The resulting finish will be rather rough and grainy and will provide a good bonding surface for the finish coat. Try to keep lap marks, ridges, and similar surface imperfections to a minimum at this stage. The coating can be from 1/4 inch to 3/8 inch thick, but should be uniform everywhere. This coat will even out any ripples and low spots in the scratch coat. As you apply the stucco, continuously check the surface for smoothness and evenness in both the horizontal and vertical planes. A long piece of straight board works well for this, and can be equipped with handles so that it can be drawn along the surface much like a screed or darby in concrete work.

Though the scratch coat can be left to cure by itself, the brown coat needs some help. Stucco cures in the same way that concrete does, by hydration. But since the sections are so thin, evaporation can easily take place so rapidly that the hydration process is adversely affected and a proper cure never does take place. As soon as the brown coat is firmly set up and curing has begun, mist the entire surface gently with the garden hose. Never use a stream, or even a hard spray, but just a gentle, fine cloud of moisture. Spray just enough to keep the surface moist, and continue doing this for at least 2 days. If the weather is hot and dry, spraying for a third day will certainly do no harm and may be of some value in the ultimate strength of the stucco. Then let the stucco air-cure for another 5 or 6 days, or a week if you wish.

Once the brown coat is well cured the finish coat can be applied. This coating may also be made from standard cements, but more often is comprised of white portland, white sand, and a coloring pigment. The coating should be at least 1/8 inch thick, and can be as heavy as 1/4 inch. Apply it with either a wood float or a metal trowel. Plaster the mix on evenly and smoothly to begin with, and administer any decorative effects you want only after a uniform coating is in place. If a smooth and unadorned finish is desired, it's only necessary to finish the surface carefully with either the wood float or steel trowel. Wood floating gives a coarse and grainy surface, while steel troweling, especially if done in two or three passes, produces an extremely smooth surface, just as with concrete. Make sure that all the corners are accurately squared with no nibs hanging out on one side or the other, and use the straightedge as you go along to insure uniformity of the surface.

Decorative effects can be accomplished concurrently with the application of the finish coat, often by a helper following behind the stucco mason. There are dozens of ways to enhance the appearance of the stucco finish. Give free rein to your imagination. For instance, you can texture the surface in many ways. A broom or brush will produce striations which can be done perfectly straight or in waves or swirls. You can apply the finish coat in wave-like ripples, and then drag a dampened wood float across only the tops of the ripples, pulling some of the material into the valleys between the ripples to create a rough, stippled surface.

Patting the surface with a moistened, flat natural sponge results in a different type of stippling. Intricate patterns can be made on the surface by using a broad-toothed comb. You can create a pleasing effect by tacking a piece of scrap carpeting to a wood float. Drawing the float across the surface gives one type of texture, while patting the surface gives an entirely different one. Carpets of different weaves generate different effects. Intricate patterns, linework or even scenes can be produced by using different texturing techniques or by alternating textures with smooth surfaces.

There is also a decorative finish which is sometimes called *rock dash*. For this you can use plain or colored glass beads or tumbled waste glass (the sharp edges are rounded by the tumbling action), white or colored marble chips, or varicolored small clean pebbles.

The process couldn't be simpler—you just stand back and throw handfuls of the material at the fresh stucco. The trick is to get even distribution and sufficient bedding of the particles. Many will not stick, so it's a good idea to lay out a sheet of plastic at the base of the project. The material which falls to the plastic can be gathered up for another shot. Larger pieces of material, which may be stone, glass or even mosaic tiles, can be set by hand, one by one. This is an extremely tedious job if the surface is large, but can be used to remarkable effect on smaller sections or accent areas.

When all the work is done, the finish coat should be allowed to set up until firm, and then moist cured with a very fine mist of water periodically applied with a garden hose, so that the surface remains continuously damp for several days. Try to keep the surface uniformly damp, and make sure that no particular area has a chance to dry out. Fresh stucco which faces the sun or is subject to hot dry winds should be protected by canvas tarps, sheets of plastic or anything else that will provide protection. These contrivances should remain in place for 4 or 5 days after completion of the project. Prevent the sheets from touching the stucco surface. Also worth noting here is the fact that it is better not to apply stucco in the direct rays of a brilliant sun. If hot sun and/or wind are unavoidable at the project site, improvise some sort of sunbreak/windbreak.

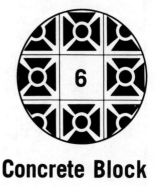

6

Concrete Block

Concrete block has grown so in popularity and availability over the past few years that it is now supplanting the venerable, kiln-dried clay brick in many types of masonry construction, and, in some instances, concrete block structures are being substituted for those traditionally fabricated from poured concrete. Concrete block has some compelling virtues; it is relatively inexpensive, very strong, lightweight, easy to handle, and purchasable in a range of styles and sizes. The variety of patterns and colors in which it is produced has given concrete block a decorative and architectural distinction all of its own. Whereas concrete block was once sequestered in foundations and commercial/industrial structures, it is now everywhere in the public eye.

STANDARD BLOCK

Standard concrete blocks are manufactured in many sizes and configurations, and are today considered the basic building module in unit masonry construction. By composition there are actually two types, *concrete blocks* and *cinder blocks*. The two terms are often used interchangeably, but shouldn't be, since the blocks differ in characteristics. Both types are sometimes referred to as *cement blocks,* but they are not. Concrete blocks are made from a standard concrete mix of cement, sand, and aggregate. Cinder blocks, on the

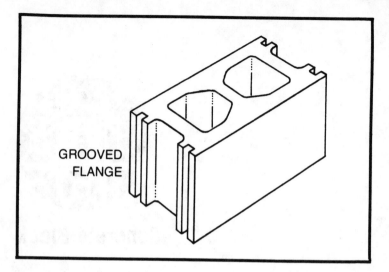

GROOVED
FLANGE

Fig. 6-1. Standard concrete stretcher block with flanged ends.

other hand, are made with cinders, slag or some similar material serving in place of stone as an aggregate. These blocks weigh only about half as much as concrete blocks, and are not as strong.

Standard concrete blocks are marketed in a range of sizes, the dimensions differing by increments of (a nominal) 4 inches. The most commonly used size is nominally 8 inches thick, 8 inches high, and 16 inches long. The actual dimensions of this block, and of the other sizes as well, are 3/8 inch less than the nominal dimensions all around. Thus a standard 8-inch by 8-inch by 16-inch block actually measures 7 5/8 inches by 7 5/8 inches by 15 5/8 inches. This provides for a 3/8-inch mortar joint at each face of the block and accords with the 4-inch module concept which has been adopted by much of the building trades industry.

One widely used size of concrete block is full height and length, but only 4 inches thick. Some blocks are 6 inches thick, others 12. Some types are available in 8-inch lengths, rather than the full 16 inches.

Within a structure each concrete block is named in accordance with its function. *Stretcher blocks* are the most numerous; these are the ones which lie end to end and comprise most of the bulk of a structure. They usually have concave ends with a flange at each side. In some stretcher block designs, the flange face is grooved, offering a key for greater strength (Fig. 6-1). *Corner blocks*, on the other hand, have one plain and squared-off end and one flanged end. *Pier blocks* have both ends plain and squared off, so that they can be stacked into a rectangular column. *Half blocks*, which are only 8 inches long, are usually squared at both ends. The thinner blocks, of either 4-inch or 6-inch thickness, are usually referred to as *partition blocks*, and are used for erecting nonload-bearing walls. There are also a number of special blocks which are molded to fulfill a particular purpose. *Jamb blocks*, for instance, are used where a door or window jamb must be fitted into place. *Sash blocks* support window sashes, *chimney blocks* are designed to fit around standard clay flue liners, and *cap blocks* are molded solid for sealing off the top of a wall or other structure. Figure 6-2 shows some of the standard styles.

Standard block may be manufactured with either two cores and three crosswebs, or three cores and four crosswebs. Which type you use makes little difference; one is just as good as the other. You will notice, too, that most (though not all) standard blocks have tapered crosswebs and face shells. Or, there may be a slight flare. The identity of the design will quickly be obvious, because one core face is notably thicker than the other (Fig. 6-3). The thicker or flared portion of the block is the top, the thinner portion the bottom. The thicker shell and web section provides a wider mortar bed surface, and should be placed upward.

The quality and characteristics of concrete blocks are expressed in a rather involved series of specifications developed by ASTM. These specifications have to do with the materials used in the manufacture, some of the methods of manufacture, application criteria, and so forth. Blocks are also classified by weight and by strength. Though there are many different sets of specifications, most of them are of interest primarily to architects and engineers. Fortunately for do-it-yourselfers, there are types (and grades) of concrete block which are formulated for general-purpose work. These are the types which most lumberyards and building supply houses keep in stock; the many other types are for specific application and more often than not are manufactured on special order only.

These general-purpose blocks are classified as load-bearing, and can be used indoors or outdoors, above grade or below grade, and are fully capable of withstanding the weather. In other words,

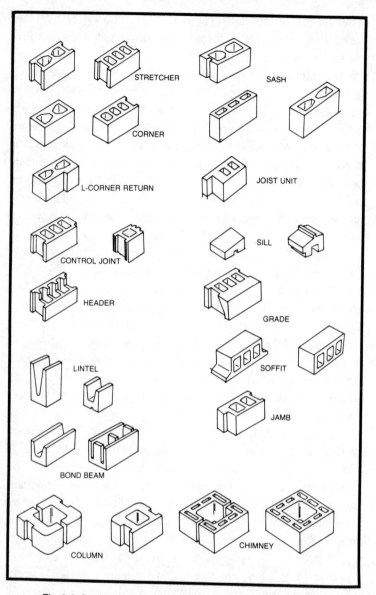

STRETCHER

SASH

CORNER

L-CORNER RETURN

JOIST UNIT

CONTROL JOINT

SILL

HEADER

GRADE

LINTEL

SOFFIT

JAMB

BOND BEAM

COLUMN

CHIMNEY

Fig. 6-2. Standard concrete blocks with tapered and flared webs.

they can be used with no problems for any project around the home. If your supplier asks you what type or grade of block you want, tell him general-purpose or utility. In locations with peculiar climatic and hydrological problems, such as an inordinate amount of frost action, a protracted freeze/thaw season, or a high water table, the dealer will most likely stock concrete blocks designed to combat these problems.

DECORATIVE BLOCK

Decorative concrete block is a fairly recent development that is now gaining a great deal of favor. Although these blocks may be selected for use because of their architectural and decorative advantages, they are also functional. Many types of decorative blocks can only be utilized in nonload-bearing applications, but others are capable of sustaining light loads. A few, particularly those which are relatively large and are either solid or contain only small cores, can be used in load-bearing applications. There are literally dozens of kinds of decorative blocks, varying widely across the country and strongly dependent upon each manufacturer's preference in design and specifications. The difference in these blocks most critical to the do-it-yourselfer, aside from the decorative design aspect, is the load-bearing capability of the particular type of block—or, more correctly—of a structural assembly of these blocks. For load-bearing applications, be sure to check with your supplier; he should have all the pertinent data on hand for the products he carries. If

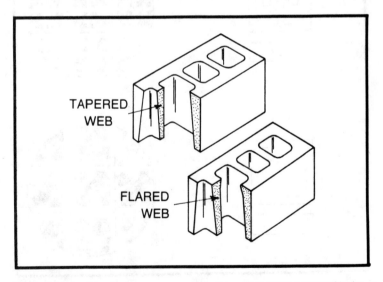

Fig. 6-3. Some of the more common configurations of standard concrete blocks.

Fig. 6-4. Representative examples of grille block.

load-bearing capabilities are of no consequence to you, then use any type of block you wish.

As far as the dimensions of these blocks are concerned, about the only constant principle you will find is conformity with the 4-inch module concept. Most block makers try to keep their products sized in 4-inch increments so that they will easily adapt to the more or less standard 2-foot and 4-foot building modules, and so that the decorative blocks can be used in conjunction with standard blocks and bricks. Occasionally you will find some odd-sized units, but these are in the minority. The most common thicknesses are 4-inch and 8-inch, but a few types are being produced in a 6-inch thickness. The faces are often 12 inches or 16 inches square, but some are only 8 inches square. Some blocks are made in long rectangles, such as 4 inches by 16 inches. All in all, the best way to begin a project built from decorative blocks is to check with your building supplier to see exactly what is available before you even sketch out the plans.

Loosely, there are several categories of decorative block, and one of the most popular is the *grille block* group (Fig. 6-4). These blocks are usually square and relatively thin, with solid faces on four sides. The cores and webs are arranged to outline a pattern in the center of the block, and there is great variety among the available patterns. There are stars, bull's-eyes, crosses, random web lines, or whatever else the manufacturer happened to settle upon. When set on edge and assembled in the form of a wall, the blocks serve as a full grille or screen which allows air to circulate, breaks up strong wind currents, and admits some sunlight although providing some degree of shade. They also afford a considerable amount of privacy, depending upon the nature of the pattern. It is difficult to see through most patterns from a distance, but easy from close up. Grille blocks are often used as low patio walls, high garden screen walls, carport side walls and the like. Though not particularly strong, grille blocks will support a light load such as a carport roof, especially when used in conjunction with an occasional supporting column built of standard block.

Screen blocks go one step further than grille blocks. They are lighter and more delicate than grille blocks, and are designed to stack face to face. They are also much more open, and the webs and faces are made about as thin as practical. The patterns molded within the

Fig. 6-5. Slump block used in lamp post construction.

faces can be quite intricate, though more often they are rather simple. These blocks will not bear a load, and are used strictly for decorative purposes. They work well as trellises for climbing plants, as high or low garden walls, and as privacy screens or windbreaks. One common application of screen blocks is in making a sort of latticework in front of large expanses of glass windows. The structure provides a considerable amount of protection for the glass, serves as a sunbreak, reduces wind pressure and turbulence, and confers a certain amount of privacy.

Slump block is a type which is both functional and decorative (Fig. 6-5). The block is solid, as a rule, and will carry loads. Slump block has excellent durability and may be manufactured to simulate large hand-finished molded bricks, adobe bricks, or rough weathered stone. Several earth-tone colors are available. The blocks have irregular faces and vary a small amount in size from block to block. This is because the mix is made in such a way that when the form is removed, the concrete slumps and settles a little bit in an uncontrolled fashion. By using two or three different nominal sizes of slump block in the same project, you can achieve an interesting random pattern effect. To further heighten its decorative effect, slump block can also be broken, split, or edge-chipped. This involves a lot of work, however, and should be restricted to small projects.

Split block has a finished face appearance similar to that of the slump block but more deeply pocked and dimpled. The split block is usually made in a 4-inch height and 16-inch length, and is laid slab-fashion. It can be laid to form a solid wall or section, or to form a pattern of openings between the units, like a lattice (Fig. 6-6). The individual units can be angled slightly to and fro, staggered, set off-center, or positioned randomly forward and back so that the faces either project or retreat. Split block can bear a sizable load and makes handsome carport or breezeway walls, low or high garden walls, patio borders and such. Like slump block, split block is available in earth-tone colors.

Patterned block (Fig. 6-7) may be solid or cored and is made in much the same manner and in the same sizes as standard block. The difference is that the faces are patterned. All sorts of patterns can be used, either recessed or raised. Normally each block is a pattern unto itself, but not always. In some cases, it takes two blocks, one

Fig. 6-6. Split block used to make a lattice wall.

atop the other, to complete a pattern. In other arrangements, four blocks are required to complete the pattern. It is even possible, by suitably placing the blocks, to make a series of small designs within a grand design.

Finished blocks start off as ordinary standard concrete blocks. Either during the molding process or after curing, one face is given a special treatment. Several different finishes are now on the market, and there will be many more in the future. Some of these finishes are smooth; some are patterned; others are vitrified or glazed. Terrazzo and marbled finishes are also available. The blocks come in assorted

Fig. 6-7. Decorative wall constructed of pattern block.

Fig. 6-8. Standard concrete block laid up with cores horizontal to make a screen wall.

colors and in a variety of textures. In most parts of the country, these finished blocks are not readily available unless you happen to live close to the manufacturer. However, many types can be special-ordered.

While on the subject of decorative block, it is worth noting that standard block can also be used in a decorative fashion, though the possibilities are more limited than with the special blocks. One common practice is to lay the blocks face to face, exposing the cores (Fig. 6-8). Pattern variations can be obtained by laying both two-core and three-core blocks (and also single-core half blocks) in a single structure. Standard block can also be set face down in the ground to serve as patios, walkways, and edging strips. When bedded in the ground with the cores facing upward, they can act as a rock garden or as a combination concrete-and-grass driveway. By stacking them at a back-sloping angle, a shallow retaining wall can be constructed that can be filled with loam and planted in grass or flowers. There are a great many possibilities, and with some imagination much can be done with standard block.

BUILDING WITH BLOCK

Most home masonry projects constructed from concrete block are walls of one sort or another. The walls are often freestanding, bearing no load except their own weight. The alignment may be straight, intersecting, or curving. Other walls are used under the weight of a light roof. The roof can be a solid one, such as covers a carport, entryway, porch or farm pumphouse, or an openwork one, such as a sun-grate over a patio or garden arbor. Concrete block can also be used for the low-wall foundation of a greenhouse or for suspending a series of cold frames. Other possibilities for home construction include piers for supporting garden benches and pedestals for supporting columns. The columns themselves might be made from concrete block, and used to buttress tall screen walls, to retain metal grillework in entryways, and to hold sheets of clear plastic around sundecks.

There are many, many possibilities for building with concrete block, and dozens of projects that you yourself can construct to enhance your property and add to its value. None of these projects is very difficult, so long as you keep a few points in mind. For instance, nonload-bearing walls that are no more than 4 feet high can be made from block as thin as 4 inches. With 6-inch units, the height could extend to 5 feet. For anything over this height, a standard 8-inch block should be used. But if you can, it is best to stay with the 8-inch size for every wall, no matter the height, since this gives a great deal more structural rigidity to the wall and renders it more resistant to shifting, cracking, and other mechanical damage.

Reinforcement

No reinforcing materials are needed in walls 4 feet high or less. The mortar joints alone will provide sufficient strength. Walls more than 4 feet high should be reinforced in one fashion or another. If a tall wall is erected from hollow-core units such as standard block, reinforcing bars should be installed in the cores. If the wall is 6 feet high, for instance, #4 reinforcing rods should be extended from the base to a height of 4 feet, and spaced approximately 4 feet apart along the whole length of the wall. The exact spacing is calculated to enable the rods to pass unobstructed through the block cores. The

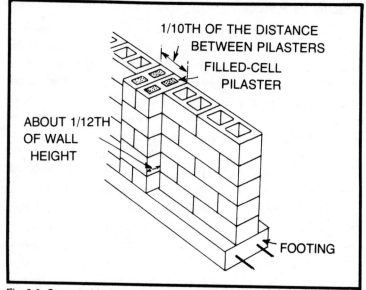

1/10TH OF THE DISTANCE
BETWEEN PILASTERS

FILLED-CELL
PILASTER

ABOUT 1/12TH
OF WALL
HEIGHT

FOOTING

Fig. 6-9. Concrete block wall construction with pilasters for extra strength and support.

cores containing the rods are filled with concrete or grout, but the cores of the top 2 feet of the wall are not. If the wall is 8 feet high, use the same method but insert two rods at each site of reinforcement, positioning them in adjacent cores and extending them to a height of 6 feet. Whether reinforcement is employed or not, face shell bedding of the mortar and a 3/8-inch joint is usually sufficient. For extra strength you can go to full bedding of the mortar.

Decorative blocks must be handled a bit differently. These are designed to be stacked face to face with the cores open to view, rather than the other way around. The situation is the same when standard blocks are stacked face to face. This means that there is no way to insert reinforcing bars. Thin units of 4-inch or 6-inch thickness can be laid 4 to 5 feet high in walls, provided that they sit on a stout, stable base. Units that are 8 inches wide or more can usually be successfully laid to a height of about 6 feet, except in locations of chronically adverse weather conditions.

Where there is a need for further structural rigidity and strength, as in cases of load-bearing walls, tall structures made from thin units, or unfavorable weather conditions, you can resort to the system shown in Fig. 6-9. The wall sections are set between masonry unit columns (pilasters) made from standard block and set every 4 feet or 6 feet or at whatever interval seems appropriate. The columns are reinforced with #4 rod and filled with concrete or grout. They can be of the same thickness as the rest of the wall, or can be thicker to provide a buttress effect. There is no need for these columns to be rectangular. You could, for instance, slant the face inward in a taper toward the top, or stair-step the lower portion, or work out some other design.

There is another method of reinforcement that can be used with concrete block construction and which will result in an exceptionally strong and sturdy structure. This is called *joint reinforcement*, and is accomplished by laying prefabricated sections of zinc-coated steel wire between the courses as the blocks are laid. This reinforcing material comes in sections about 10 feet long and is manufactured in two styles; the *ladder* type, and the *truss* type (Fig. 6-10). Both will do an equally good job. Joint reinforcing wire is placed in the fresh mortar bed so that the wire is completely surrounded by mortar. The wire is laid in strips for the full length of the structure. Adjacent strips should be overlapped by at least 6 inches, and special fabrications used for laying the wire around right-angle corners. For other angles the wire must be bent to suit as the occasion arises.

Usually there is no necessity to use joint reinforcement on every course of block. Every second course greatly increases strength, and every third course is more than adequate for most jobs. The joint thickness remains 3/8 inch. The reinforcement,

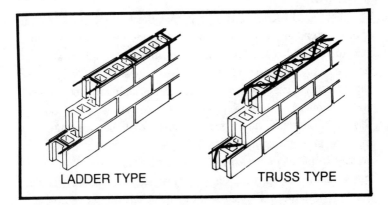

LADDER TYPE   TRUSS TYPE

Fig. 6-10. Reinforcement used to strengthen concrete block wall.

which comes in several widths, should be accurately positioned on the blocks so that the distance from the outside face of the structure to the reinforcement will allow for a minimum 5/8-inch cover of mortar. The distance from the reinforcement wire to the inside face should allow for a minimum 1/2-inch mortar covering. Joint reinforcement can be used in standard block construction where the cores run vertically, and also where the blocks are assembled face to face with the cores horizontal. It can be placed in solid block structures as well, though this is not usually necessary, especially if full-bed mortaring has been incorporated.

Footings

Whatever the design specifics of the block wall or other structure, the whole affair must rest on a solid base. This means that the starting point is a poured concrete footing. Without such support, building a plumb and level structure is difficult, and subsequent settling, shifting, and cracking is bound to take place. In fact, there is little likelihood that the wall will remain intact for any great length of time.

Footings for these structures are built in the usual manner, either poured directly into a trench, or into forms. The earth beneath the footing should be thoroughly tamped and compacted, and a layer of sand or gravel added for drainage if necessary. The top of the footing needs no keys, but should be finished true and relatively smooth, as free as possible from humps and bumps, and level in all directions.

The usual recommendation is that the bottom of the footing should extend at least 18 inches below grade level. This is a good rule to follow, because that depth is sufficient to afford an extra degree of stability and solidity by virtue of the soil tamped in around the structure's base, which in most cases includes the footing itself and a couple of courses of masonry. In many areas of the country, this depth is sufficient to guard against frost damage. Alternate freezing and thawing and the heaving action of frost in the ground creates tremendous pressures. Over a period of time, these pressures can wreak havoc with even the sturdiest of freestanding masonry structures. If you live in an area of hard winters and severe frost action, lower the base of your footings accordingly so that they are below the average frostline. If you are not sure of the local frostline depth, check with professional building contractors or excavators in your area—they'll know. In some few areas, this rule is disregarded simply because of the impracticalities involved. There are places where the frostline can easily descend to 5 feet—even deeper during particularly severe winters. Footings, however, are seldom placed lower than 3 1/2 or 4 feet, although in compensation they may be constructed wider than normal and more heavily reinforced.

The footings must be capable of supporting a considerable amount of weight (even a relatively low wall structure is heavy). Most likely they'll be subjected to a certain amount of stress and strain, too. This means that setting plenty of reinforcement into the footings is a good idea. For most applications, a pair of #3 rods for small footings and #4 rods for larger ones will be ample. The rods should lie lengthwise along the footing, one under each wall face at half the depth of the footing (Fig. 6-11). If reinforced columns are to be set atop the footings, or if the wall itself is to be reinforced, then additional bars should be bent up out of the footing at appropriate points. If you feel that frost may be a problem, or just want an extra-sturdy footing, add more reinforcing bars.

LAYING OUT BLOCK STRUCTURES

The length of the footing should be figured to protrude beyond the ends of the block structure by the same distance that it extends beyond either side. The overall length of the structure should be sized to the modular dimensions of the blocks. In other words,

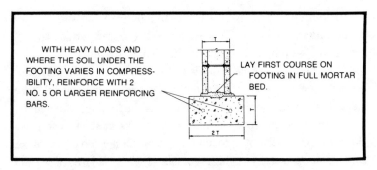

WITH HEAVY LOADS AND WHERE THE SOIL UNDER THE FOOTING VARIES IN COMPRESSIBILITY, REINFORCE WITH 2 NO. 5 OR LARGER REINFORCING BARS.

LAY FIRST COURSE ON FOOTING IN FULL MORTAR BED.

Fig. 6-11. Standard footing arrangement for concrete block wall.

instead of choosing arbitrary dimensions for the structure, the dimensions should be made to conform with the length of an individual block. This makes for a better structure, and precludes the additional labor involved in cutting a series of blocks.

A 10-foot wall, for instance, would require 7 1/2 standard 16-inch blocks on every course. This would work out fine, because you can use seven full blocks and one half-block in each course. However, if you elected to make an 11-foot wall, then you would need 8 1/4 standard 16-inch blocks per course and a lot of cutting would be entailed. Of course, if you are locked into a particular dimension, then you must adapt the blocks to suit the constraints. But if you have free choice, let the size of the units determine the overall dimensions.

The same situation can arise where an opening is needed in a block structure. Suppose, for instance, that you sketch out plans for a decorative screen wall with an arched gateway to stand along the property line in front of your home. If you make the overall length of the structure 42 feet 3 3/4 inches you'll have to do a lot of cutting. If you position the gateway 19 feet 7 inches from one end, you will wind up with some strange-looking corners and ends, and the joints will not be symmetrical. By the same token, if the gateway is 43 inches wide, you will have difficulties with both wall sections and the archway as well. So start with the rough overall dimensions that seem best suited to the project, but then either expand or compress them slightly to accommodate the 4-inch module increment principle.

LAYING BLOCK

The first step in laying block is to look over the site and get everything in readiness. We'll assume that the preliminary work of laying out the project and making the footings is done. Assemble handy to the work site all the tools and equipment that you will need. Clear any debris or piles of dirt out of the way so that you will have plenty of working room the whole length of the footing, with nothing lying in wait to trip you. Distribute the blocks in stacks and rows along the full length of the footing, close by but not too close. Make sure that the top of the footing is completely clean and free of dirt and dust by giving it a thorough sweeping.

Location and Alignment

The next step is to establish the exact position of the first course of blocks. The lengthwise centerline of the course should lie directly on the lengthwise centerline of the footing. Find the centerline of the footing, measure out to one side or the other one-half the thickness of the blocks you are using, and snap a chalk line the full length of the footing. This will serve as the guide mark for the face of the first course. Measure from this line to the outside edge of the footing, and make a mental note of the figure. If the guide mark should disappear under some mortar or be rubbed out inadvertently, you can check the block alignment by measuring in from the edge of the footing.

When starting a first course, most masons will set in place all of that course's blocks, properly aligned and spaced but without any mortar applied. This serves as a doublecheck to make sure before any blocks are laid that the basic dimensions are correct. This is a good idea, and well worth doing. As soon as everything is verified, you can start laying up.

Mortar Considerations

The next task is to mix up a batch of mortar. Though this could have been done earlier, it's a good idea not to do any mixing until you are sure that everything is ready to go. This avoids the problem of having mortar start to stiffen up in the tub before you are ready to use it, in the event that the preparations take more time than expected.

About the best mortar to use for home projects is a standard Type M, which calls for 1 part portland cement, 1/4 part hydrated lime, and 3 parts sand. If you are using masonry cement, use 1 part portland cement, 1 part masonry cement Type II, and 6 parts sand. Depending upon the nature of the sand, the portland cement mix can require as little as 2 1/4 parts, and the masonry cement mortar as little as 4 1/2 parts; this is variable according to specific circumstances. If conditions are favorable where the project is being constructed—that is, no severe frosts, strong winds, and no heavy loading—you can drop back to the leaner Type S mortar. A Type N could also be used, but only with solid masonry units or those which are stacked face to face; it should not be used with hollow units.

Fig. 6-12. Spreading full mortar bed on footing top for starter course of concrete block.

Keep the Blocks Dry

Another point worth noting here is that the blocks should be absolutely dry. They should never be sprayed with a hose in an effort to make a better mortar joint, as is sometimes done misguidedly, because exactly the opposite will happen. Concrete blocks expand with moisture, and if laid wet, they will eventually dry out and shrink, destroying the joint bond.

Bedding the Mortar

To start the proceedings, throw a full bed of mortar on the footing, working from the corner or end back along the footing surface. Make this ridge of mortar long enough to accommodate three to five blocks. Spread the ridge of mortar by making a furrow in the center of the ridge and curling the mortar outward to each side in a fairly even strip (Fig. 6-12). It is difficult to explain just how much mortar should be used, and this is something that you will discover with experience. However, it is better to lay on too much than too little. Remember that the joint must have a finished 3/8-inch thickness. The width of the strip of mortar should be just slightly less than the width of the blocks, so that the edge of the mortar runs along just inside the chalked guideline.

Laying the First Block

The first block is a *corner block*, regardless of whether it is actually positioned in a corner or will become an end block. Position this block and lay it gently on the mortar. The best way to handle a block is to pick it up from the top, gripping an end web and a center web with each hand so that the block hangs downward. You can then lower the block into place while siting it with your eyes at the same time.

After the block is bedded in the mortar, quickly make all the necessary adjustments. First make sure that it is precisely aligned with the guideline. Push the block down firmly so that mortar squeezes out in all directions and the joint reaches its proper 3/8-inch thickness. You can determine this by taking a quick measurement with a tape; each upper corner of the block should rise exactly 8 inches above the footing top. Lay a spirit level lengthwise across the top of the block, and check for level. Tap the block very gently with the trowel handle as necessary until the block is dead level.

Stand the level up along the face of the block and check the plumb. Again tap the block gently with the trowel handle to bring the block into plumb if necessary. Then check the level again. You may have to repeat this process two or three times if slight adjustments have to be made from one direction to the other. But don't leave it unless you are satisfied that the corner block is absolutely level and absolutely plumb and in perfect alignment. If it is not, all sorts of difficulties will be encountered later on.

As you are making the adjustments, take care that the block does not actually break loose from the mortar. It should remain completely embedded, and never be picked up or moved any great amount after the mortar has been fully pressed down and squeezed out to form the joint. If you think that the joint is not a good one and that the bond may have broken while you were fiddling with the block, take up the block, lay some fresh mortar, and begin again.

Laying the Rest of the First Course

As soon as the corner block is squared away, start laying the *stretcher blocks*. Set on end as many blocks as you have laid mortar for; set each block close to the spot it will occupy in the wall. Butter mortar onto the shell or flange of each vertical face (Fig. 6-13). Pick the first block up and suspend it over the mortar bed roughly in position, but an inch or so back from the already laid corner block, so that the fresh mortar of the two vertical joints does not come into

Fig. 6-13. Buttering the end flanges of concrete stretcher blocks preparatory to laying them.

contact (Fig. 6-14). Keeping the block in correct alignment, lower it into the mortar bed and at the same time move the vertical flanges together. This is done in a single diagonal movement, smoothly, and the movement should end as the block comes to rest when the mortar in the vertical joints has compressed to a 3/8-inch thickness, and the mortar in the bed has squeezed out to a 3/8-inch thickness.

Fig. 6-14. Laying first-course stretcher block in full mortar bed.

The technique should not consist of a vertical descent and a horizontal coasting into the already laid block, as this will not result in a good joint.

Now you have a choice to make. You can go ahead and level and plumb the single block you have just laid, or you can continue to lay the two or three remaining blocks in that set. The latter action is generally preferred, because it is a bit easier to line up, plumb and level several blocks at a time than it is to work with individual ones. The best tool for this purpose, by the way, is a long spirit level which will extend over at least three standard blocks (48 inches).

Assuming you choose to work with the set of blocks, first check for level, including the corner block previously laid. Tap the webs with the trowel handle until all of the blocks are in perfect level (Fig. 6-15). Then align the blocks by using the level as a straightedge along the block faces, both on the inside and on the outside of the structure. Check each block individually for plumb and bring it into line as necessary (Fig. 6-16). Remember that as one type of adjustment is made another may be thrown out of kilter, so doublecheck everything as you go along. The importance of getting this first course absolutely true cannot be overemphasized, as it forms the basis for the entire structure. If mistakes are made here, you will

Fig. 6-15. Checking the first set of blocks for level.

Fig. 6-16. Checking the first set of blocks for plumb.

have to attempt to correct them as the construction proceeds, and this is inevitably a good deal less than satisfactory.

After the set of blocks is completely trued up, scrape away the excess mortar that has extruded from the joints and flip the scraps back onto the mortar board for reuse. The mortar that has curled out from the full bed on the footing may be left in place if the first course will later be covered with fill dirt, or, it can also be scraped away and reused.

Corners and Ends

The first set of blocks are now in place, and the next step depends upon the nature of the project. If it is a straight wall, go to the opposite end and do the same thing all over again, laying three or four blocks of the starter course. If you are working at a corner, the next set of blocks is laid from your original starting point, and run off at a 90-degree angle. In this case, the procedure is just the same, except that the first block you lay will not be a corner block, but a stretcher, since it will butt up against the inside face of the original corner block (Fig. 6-17). Lay this block gently and with closest attention, making sure that you do not dislodge the first corner block. In fact, take great care not to bump any of the blocks, as it doesn't take much to ruin a joint before the mortar begins to set up hard. Again, line up this next set of blocks perfectly.

If you have just laid out a corner, then presumably the two wall sections will either have ends or will form other corners. The next step is to lay the starter course—again in sets of three or four blocks—at these ends or corners until all are done. Then, despite the gaps left between the block sets, start to lay the second course.

Go back to the original starting point, and lay a face shell mortar bed on top of the already laid blocks. Mortar should also be thrown across the end web of the corner block. The length of the mortar bed should extend almost the full length of the laid set of blocks. The second course will be a half-block shorter in length (Fig. 6-18).

If you are working at a corner, the mortar bed should be thrown on the first half of the original corner block and then along the set of blocks that you laid second, rather than on the first set. This is done so that the blocks will interleave at the corner, one block heading in one direction, the next in the other, and so forth.

Set the first block carefully in place atop the first course. If you're at the end of a wall, you will use a half block. Then the following stretcher blocks will be staggered over the joints in the first course. If you are working at a corner, the first block will be a corner block. Align, plumb and level this first block carefully, just as you did on the first course. Then stand the stretcher blocks on end,

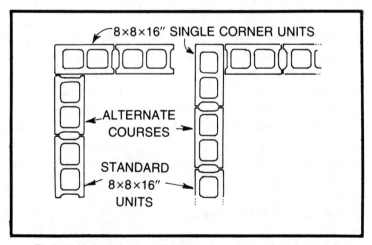

Fig. 6-17. Method of setting alternate corner courses of concrete block.

Fig. 6-18. Standard concrete blocks arranged in a stepped running bond construction, one-half overlap.

butter the flanges, and lay them in place. When you finish this set, you should be a half block short of the end of the first course. Align, level and plumb the set of blocks, and scrape off the excess mortar.

Now continue to build the ends or corners up pyramid-fashion, until you can go no further. With a simple wall, this will mean only building up the two ends. With corners, you will be working in both directions (Fig. 6-19). In either case, these sections will be four or five courses high when you finish, depending on the number of blocks you originally laid in the starter course. When you build up the corners and ends, check them frequently with a long spirit level to make absolutely sure that they are plumb. Also, doublecheck the block spacing by placing a straightedge or level diagonally across the corners of the blocks (Fig. 6-20). If the spacing is correct, all the block corners will line up exactly. If they do not, you have made an error somewhere which will either have to be corrected now or kept in mind so that you can compensate for it later.

Completing the Starter Course

The next step is to finish up the starter course between the sets which were originally laid. Run a mason's line out taut from corner to

Fig. 6-19. Corners are built up first in block construction.

corner or end to end, lined up exactly with the line of the top edge of the course. You can do this by using special holders which attach to the existing blocks in the course, or you can just secure the line by laying a block on each of its ends. However you do it, make sure that the line is tight and straight, with no sag in the middle. This will serve

Fig. 6-20. Checking the diagonal line at corner to make sure that all blocks are properly aligned.

225

Fig. 6-21. Applying mortar to all sides of the closure opening.

as the guideline as you lay the course of block. Each block should align with the mason's line exactly. When this is done, start laying the stretcher blocks in place three or four at a time, just as you did before. On this first course, use a full mortar bed. You can work from one end toward the other, or work from both ends toward the middle; this makes no difference.

Sooner or later you will come to the point where only one more block is needed to complete the course. This block is called a *closure block*, and is installed just a little bit differently than the others. To begin with, the gap in the course must exactly match the size of the block plus the mortar joints at each end. If you have set the stretcher blocks carefully and maintained the correct vertical joint thickness, the space should be exactly right for the closure block. Check this first so that you will know where you stand. Assuming the fit is all right, butter all four vertical edges or flanges of the blocks already laid that form the opening. Lay a face shell bed of mortar on the course below in the usual fashion (Fig. 6-21). Then butter all of the vertical edges or flanges of the closure block. Suspend the block above the opening, accurately aligned, and slip it slowly and gently down into the gap (Fig. 6-22). The vertical mortar ridges should blend and slide together as the block moves downward. You'll have to be careful not to dislodge the mortar. Press the block down into the mortar bed to form the usual bed joint. Align the block, check for

Fig. 6-22. Slipping closure block in place.

level and plumb, and scrape the excess mortar from the joints. That completes the course.

The Succeeding Courses

Remove the mason's line from the top of the first course, and move it up to the top of the second course. Repeat the whole procedure (Fig. 6-23) to complete the second course, and move on

Fig. 6-23. Laying second-course stretcher blocks between corners.

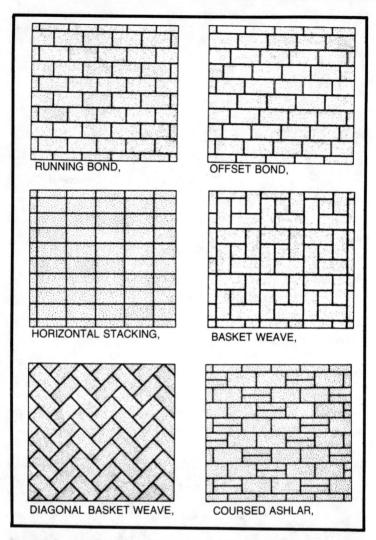

RUNNING BOND,

OFFSET BOND,

HORIZONTAL STACKING,

BASKET WEAVE,

DIAGONAL BASKET WEAVE,

COURSED ASHLAR,

Fig. 6-24. Some of the more common bonds used in concrete block construction.

to the third. If the whole height of the wall is the same as the corners or ends that you have already done, just continue to fill in the courses until the structure is complete. But if the wall is to be higher, continue to build up the corners or ends as the job progresses. These instructions have assumed the laying pattern to be a running bond. Slight procedural modifications are necessary if other bonds (Fig. 6-24) are laid up.

Topping the Structure

If the wall is laid up with solid masonry units, or with cored units which are laid face to face with the cores set horizontally as in a screen wall, then no further treatment is necessary. The top of the wall presents an unbroken surface, and if the joints are properly made, no moisture can work its way into the structure. But a hollow-core wall laid with the cores in the vertical position must be topped with solid units. Otherwise, moisture from rain and snow will seep down into the cores and cause problems. The usual method of topping a wall is to run a course of *cap block*, 4 inches thick (Fig. 6-25). There are many other possibilities, though, such as precast concrete slabs or specially cast coping (Fig. 6-26), both of which cover two or three blocks at a time and may be either of the same width as the wall, or wider to provide an overhang. Cut slate could be used, or some other type of stone (either dressed or not), provided that the units are solidly mortared in place. Cores may also be filled with mortar or concrete.

JOINT FINISHING

As work on the wall proceeds, take time out periodically to work the joints, if the plans call for it. Sometimes no further work is needed, as with an extruded joint. This type of joint is occasionally used in decorative walls, and involves merely leaving the squeezed-out mortar completely alone as the blocks are laid. Sometimes an

Fig. 6-25. Finishing off a concrete block wall with a course of solid cap block.

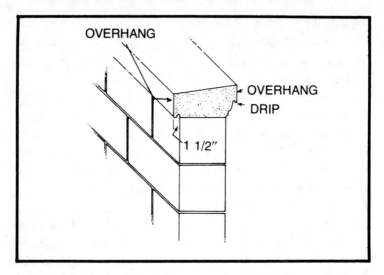

Fig. 6-26. One of many styles of precast concrete coping used to cap the top of a hollow-core concrete block structure.

extra amount of mortar is thrown, so that a substantial curl of excess mortar will remain after the block is set. Flush joints are sometimes used, too, although this is more common with interior walls or those which are protected from the weather. The flush joint is accomplished when the excess mortar is scraped away from the joint as the block is laid. No further attention is needed, and the joint is just left as is. However, concave or vee-joint treatments are usually preferred for exterior walls subject to weathering. These joints have a lower susceptibility to moisture penetration than most other types. Raked joints are also satisfactory, but this type can be more readily penetrated by moisture.

When to Tool the Joint

If joint tooling is to be done, it should be attended to after every few dozen blocks have been laid, or about every hour or so, depending upon weather conditions. The tooling should be done after the mortar has had time to set up a bit, but while it is still moist and workable enough to take the proper shape. If you can press down on a joint with your thumb and leave a thumbprint, the time is right. As the joint is tooled, the mortar is compressed and made denser, and thus more weather-resistant, at the surface, and at the same time rammed down tight along the edges of the block to make a complete seal.

How to Tool The Joint

Tooling joints is a simple process that can be learned in short order. Do the long horizontal joints first with a tool of appropriate face shape (round bar, square bar held at an angle, etc.) and of about a 2-foot length. The long length will give you a smooth and chatter-free finish. The width of the tool should be just slightly greater than the width of the joint, so that a good seal along the edges of the block can be made. Run the tool along the joint in long sweeps, pressing inward as you do so (Fig. 6-27). When all of the horizontal joints of a particular section have been completed, go back and tackle the vertical joints with a small S-shaped or angled jointing tool of the same face shape as the large tool. Work these joints carefully and feather the joint ends into the horizontal joints in a smooth transition. The tooling will probably leave a series of small burrs and particles of excess mortar clinging to the block at the outer edges of the joints. Run the edge of a small trowel along the block faces to remove these dregs, but be careful not to stab the trowel into a finished joint.

CLEANUP

At various times as the construction proceeds, take a moment off to inspect both faces of the completed work. You may find that

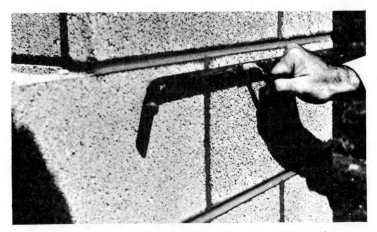

Fig. 6-27. Tooling horizontal joints with a long jointing tool.

some gobs of stray mortar have stuck to the block here and there. Now is the easiest time to clean them off with the blade of a trowel, before they solidify. After the job is complete, check all the surfaces again for imperfections or missed mortar droppings. Fill in pockmarks or other defects with a small amount of fresh mortar, carefully applied so as not to smear the mortar. You can even use a putty knife for this purpose.

Dried particles or clumps of mortar can be removed with a putty knife or a chisel, or scrubbed off with a chunk of concrete block. Mortar which has adhered to the block faces will leave obvious blemishes after drying out completely. These stains should be scrubbed with a chunk of block first, and then vigorously wire-brushed to remove as much of the mortar as possible. Take as much care as you can during the construction process to avoid mortar-spotting and staining, because under some circumstances these blemishes are difficult to remove and can be most unsightly. This is especially true when a mortar of contrasting color is involved, or when stains appear on colored block. Sometimes these stains are impossible to remove, and the various commercial masonry cleaning solutions worsen the problem instead of correcting it.

INTERSECTING WALLS

If you are building intersecting walls, such as where one wall will join another to form a tee, there are two procedures to follow depending upon whether the structure is to be load-bearing or not. The intersecting wall is not constructed as an integral part of the crosswall; instead, it is left freestanding and connected to the crosswall with metal ties. Though there is a masonry bond between the two walls, the methods of joining them allow for a certain amount of flexibility and slight movement (as from expansion and contraction), so that the walls will not crack open.

Ties

In nonload-bearing cases, the ties consist of pieces of metal lath (the same as is used in plastering or stuccowork) or 1/4-inch galvanized hardware cloth (a type of screening) sized a bit narrower than the width of the masonry unit and of about the same length. When one wall is built before the other, a tie is placed atop every

Fig. 6-28. Nonload-bearing partition wall is tied to crosswall with a section of hardware cloth embedded in mortar joint.

second course at the point of intersection, with half of the tie embedded in the mortar joint and half sticking out. When the intersecting wall is constructed the exposed portions of the ties are mortared into the joints as the job goes along. If both walls are built together, the ties are mortared in as the joints are made during the course of construction—again, at every other course only (Fig. 6-28).

Tiebars

If the walls are load-bearing, the crosswall and the intersecting wall are built at the same time, and heavy metal tiebars are used to join the two walls. These tiebars are strips of 1/4-inch-thick metal, 29 inches long for standard block and 1 1/4 inches wide. The ends are bent at opposing right angles, one end up and the other end down. The tiebar is laid across the intersecting wall and into the crosswall, with the crosswall angle pointed up into a core space, and the intersecting wall angle pointing down into a core space. The cores are filled with mortar, with wire screening placed at the core bottoms so that the mortar cannot drop through (Fig. 6-29). These tiebars are spaced vertically at 4-foot intervals within the walls.

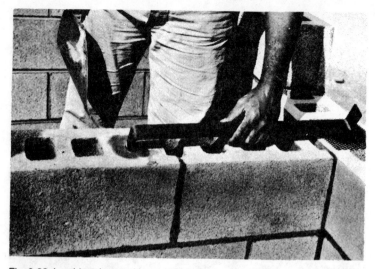

Fig. 6-29. Load-bearing partition wall is tied to crosswall with heavy iron or steel tiebar mortared into block cores and bedded in mortar joint.

The Control Joint

The vertical flanges or edges of the blocks at the end of the intersecting wall are mortared in the usual fashion and set against the crosswall to make a joint of the same thickness as the rest of the wall joints. This particular type of joint is known as a *control joint*. If the walls won't be exposed to the weather, this joint is finished to match the other joints. However, if these are exterior walls the joint should be raked out to a 3/4-inch depth after the mortar has set up quite firm. Clean out the joint opening thoroughly, and fill the space with a good grade of construction caulking. Force the caulk into place with a trowel if it is of a stiff consistency, so that the joint is completely full. The more free-flowing types can be put in with a caulking gun, but make sure that there are no voids or large air bubbles. If the appearance of the finished joint is important, the caulking should be of the same color as the mortar, and care should be taken not to smear the caulk on the block faces, where it is nearly impossible to remove.

WALL OPENINGS

Full-height openings in a wall section can be handled in a couple of ways. The wall can be treated as two separate sections, each with

234

its own footing and each individually built, with the two parts being set in alignment during the construction process. The wall ends are treated in the usual fashion. Or, the two wall sections can be individually built on one full-length footing, with an appropriate space left between the wall sections for the opening. The top of the footing at the opening can simply be buried with fill dirt, or brought up almost to grade level and then capped with a *sill* of stone or precast masonry. There are also a number of other sill treatments which could be used. The full-length footing, whether used with a sill or not, is a good idea, especially if there will be a gate in the opening. The footing will help keep the two wall sections in alignment, whereas in a wall of two separate sections, one section may tilt in one direction and the other in the opposite, from ground settling or frost heaving. This would naturally throw a gate or other closure out of alignment.

Lintels

Openings within a wall section, whether for doors, gateways, windows or ports, can be arranged in a number of ways. In all cases the wall section immediately above the opening must be supported by a *lintel* (Fig. 6-30). Lintels take a great many forms, and may

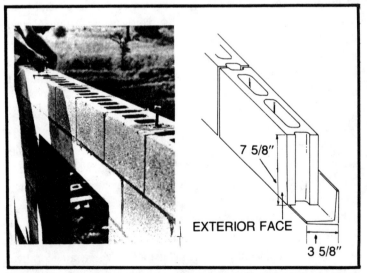

Fig. 6-30. Precast concrete lintel over wall opening (left), method of using heavy angle iron as lintel (right).

Fig. 6-31. Precast concrete sill being placed in block wall opening after wall is built up.

consist of a full-length precast reinforced section of concrete, a full-length section of cut and dressed stone, a series of individual stone blocks, or courses of block, brick, or other masonry building units. All but the reinforced concrete or full-length stone lintels must be further supported by heavy angle iron at both front and back faces, or by a special steel lintel. Whatever is used must have great strength, so as to tie the wall sections together across the opening, and at the same time support the weight of the masonry courses above. Lintels should extend into the side walls a minimum of 4 inches and preferably more.

Sills

Sills are used to cap the masonry course which forms the bottom of the opening (Fig. 6-31). These are also available in reinforced precast concrete, or can be made up from special masonry units or from dressed and fitted stone. Most sills overhang the face of the wall by 1 to 2 inches, and are slanted to drain moisture away. In some cases, however, the sills may be fitted flush with the wall faces. This might be the case, for instance, where a decorative grillwork is set into a patio wall section or stood as a freestanding structure in a garden. Though the lintel is still used, the sill can be capped in any practical fashion and the sides of the opening consist of the masonry unit construction itself. The metal grille can be bolted or mortared into place, depending upon its design.

Lug sills extend into the masonry wall on each side to an extent of about 4 inches, and sometimes more. These are installed as the construction of the wall section proceeds. *Slip sills*, on the other hand, fit between the sides of the opening and are installed after the wall itself is completed. Where the openings are intended to be filled with windows or doors, sills and lintels may be specially cast to receive the window or door frames, and special jamb blocks, or bricks, fitted in a jamb configuration to form the sides of the opening. In some types of construction, frames may be installed first and the masonry built up around them, while in others, the frames are installed after the masonry construction is complete.

PIERS

Piers can be constructed quickly and easily from concrete block, and the method of construction is much the same as for a wall. Regardless of the size of the pier, it is essential that the structure be perfectly plumb, especially if it is to bear any significant amount of weight. The masonry units in an out-of-plumb pier structure that is heavily loaded are subject to a lot of peculiar stresses and strains and can eventually fail.

About the smallest practical pier is made from a series of half blocks, each measuring 8 inches × 8 inches × 8 inches, either solid or having a single core. These units are stacked one on top of another with a face shell mortar joint for hollow-core blocks and a full mortar bed joint for solid ones. Hollow-core piers can be further beefed up by inserting a couple of reinforcing rods and pouring the core full of grout.

Larger piers can be made from standard 8 inch × 8 inch × 16 inch pier blocks, which are squared at each end. Build a suitable footing first, then place the first two blocks side by side on the footing, in a full mortar bed at the bottom and with either a full mortar bed or a face shell bed between the two blocks. The next pair of blocks is set at right angles to the first, the third set at right angles to the second, and so on. The cores should be placed vertically, and reinforcing rods set into the cores and embedded in grout.

If the piers are not load-bearing, reinforcement is usually not necessary. Nonload-bearing piers which primarily serve in a decorative or architectural function can be made with horizontally-laid cores or even with screen block or some other kind of ornamental block. Heavy reinforcement is not used, and the blocks can be mortared together in whatever fashion best fulfills the decorative purpose and at the same time affords sufficient bonding surface to hold the units together. Joint reinforcing is often a practical means of providing extra strength.

ANCHORING

The addition of a superstructure, most frequently made of wood, to the top of a wall section or pier requires the setting of some sort of anchors to hold the superstructure in place. This would be the case, for instance, when a roof is attached to a carport, or a roof or sun-grid is supported above a patio. The same principle is used in securing a deck frame to piers or columns, mounting lamps or grillwork in or on the masonry, or in similar circumstances.

Anchoring the Top Plate

Setting anchors in hollow-core blocks laid with the cores vertical is a matter of embedding the proper anchor in an empty core. This presumes that the top of the wall will be capped with a continuous wood plate. The best method is to set a piece of screening or metal lath across the core opening two courses down from the plate location, to keep the mortar from dropping through. A long bolt is stood head down on the screen, and the core is filled with mortar. As the top course is laid, the protruding bolt is again bedded in mortar. Running through two courses gives the anchoring point plenty of strength, and is the best method for load-bearing applications (Fig. 6-32).

When the top of the wall will be exposed to the weather, it must be capped with solid masonry units to keep the weather out. In this case, the anchor bolt can be brought up either through the top two courses of block and then through the cap course, or, if the intended load is relatively light, through only the top course of block and the cap course. The easiest way to bring the bolt out is to figure the location ahead of time so that the bolt will lie directly in the middle of a

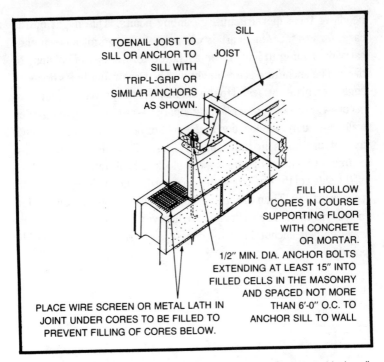

TOENAIL JOIST TO
SILL OR ANCHOR TO
SILL WITH
TRIP-L-GRIP OR
SIMILAR ANCHORS
AS SHOWN.

JOIST

FILL HOLLOW
CORES IN COURSE
SUPPORTING FLOOR
WITH CONCRETE
OR MORTAR.

1/2" MIN. DIA. ANCHOR BOLTS
EXTENDING AT LEAST 15" INTO
FILLED CELLS IN THE MASONRY
AND SPACED NOT MORE
THAN 6'-0" O.C. TO
ANCHOR SILL TO WALL

PLACE WIRE SCREEN OR METAL LATH IN
JOINT UNDER CORES TO BE FILLED TO
PREVENT FILLING OF CORES BELOW.

Fig. 6-32. One method of anchoring plate to top course of concrete block wall.

cap block head joint. The ends of the abutting cap blocks may have to be gouged away slightly to clear the bolt, and the resulting gap filled with mortar.

Where anchor bolts are used to attach long plates, the spacing should be about every 4 feet, and the bolts of 1/2-inch diameter. However, this depends to some degree upon the loading factors. Remember that the load consists not only of the weight of the superstructure itself, but also of any snow load which might accumulate, as well as the load imposed by wind. Such loads can easily combine to a total of 150 pounds per square foot or more, but could be as little as 50 pounds per square foot, or even less. For light loads, the anchor bolt spacing can be increased, and/or the bolt diameter decreased.

Other Anchoring Considerations

Where a long and continuous plate is not used, the anchor bolts and their specific locations must be chosen to match the particular needs of the superstructure mounting points. This might be the case, for instance, in an arbor where the exposed main beam-ends carrying a sun-grid have to be anchored to the tops of columns or piers. The anchoring points might well be two or four holes drilled in angle-iron plates attached to the beams. Or, the problem might be to secure a series of outdoor lamp standard bases to the top of a garden wall. Obviously, the location of the anchor bolts must exactly match that of the holes in the superstructure. Sometimes the easiest course is to omit all or part of a cap block or final course block, close off the sides of the opening with form boards, and fill the opening with a concrete mix. Then the anchor bolts can be laid out and embedded exactly without much difficulty. Whatever the circumstances, always embed anchor bolts as deeply and firmly as is reasonably possible.

Brick

Brick in one or another of its many forms is an all-time favorite masonry unit, and is second only to wood as a basic construction material for structures of all kinds. The first bricks were of the sun-dried or adobe type, and so far as we know were in use nearly 4,000 years before the coming of Christ. Early civilizations learned how to make burnt or kiln-fired brick, both plain and glazed (enameled). Though stone was probably the most widely used building material through the centuries, brick manufacture became increasingly widespread and brick construction more and more common. The first documented brick construction in this country was a residence built in New York in 1633 of bricks imported from Holland. The manufacture of bricks started in the Colonies somewhere around 1650 and today is a tremendous industry that produces hundreds of brick styles and millions of units for all manner of purposes.

COMPOSITION

All bricks are made of clay, and are divided into two rough categories according to the composition of the clay from which they are manufactured. One category is produced from clays or shales which contain only a small amount of carbonate of lime. The second category is produced from clays which contain a large amount of carbonate of lime. Hydrated aluminum silicate is the principal ingredient of the first category of clay, and tends to make the strongest brick. The second category contains a great deal of marl, a soil deposit used especially as a fertilizer.

COLOR

When these clays, and there are many, many slightly differing varieties to be found all over the world in surface deposits, are crushed and mixed with water preparatory to the brick manufacture, they look pretty much the same. Their consistency is approximately the same, and their colors range through various nondescript shades of gray earth-tones, just as does any wet clay that you might see in its natural state. But once that clay is shaped and baked to make brick, the colors appear. There is no way to tell beforehand just what the color will be, except by previous experience and analysis of particular clay deposits. The principal coloring agent in the clay is iron oxide, and the percentage of that material in the clay mix, along with minor contributions by some of the other ingredients, determines the final color of the fired brick.

MANUFACTURE OF BRICKS

When the manufacturing process begins, the raw material consists essentially of clay and water. Any appreciable quantities of organic material will render a clay useless for brick-making, and other impurities such as chlorides and potash are either removed or chemically neutralized. The mix is stiff but plastic, much like a thick mortar, when the heat treatment begins. This is a precise and critical operation, because if the material is simply heated for a while, the water is slowly evaporated away, and since no significant change has taken place, the material is crumbled for reuse or discarded. The material has not been transformed, the clay has merely dried out. But with rising temperature and extended baking time, the particulars of which must be just right, the molecular structure of the material changes and the material solidifies. If the heating is continued, the particles will begin to fuse, and eventually will join together into a solid mass. Soon afterwards the melting point is reached and the material changes character completely.

Handmade Bricks

Until recent times, brick was always made by hand, and the old handmade brick is much in demand today—for restoration work and also because of its distinctive characteristics. Handmade brick is inevitably somewhat irregular in shape and not particularly uniform in overall size, and has interesting color variations from type to type and from brick to brick. Some brick is still handmade today in small quantities, and in fact, you can make brick yourself if you are prepared for a great deal of hard work and have an ample supply of oak or beech cordwood on hand. The process is tricky, demanding, and requires a tremendous amount of hard work, but is also an exciting and interesting challenge for a dyed-in-the-wool do-it-yourselfer.

Manufacturing the Modern Brick

Most of us, though, prefer to buy our brick ready to go from the nearest supplier. Modern manufacturing methods have given us an abundant assortment of brick types, styles, and colors from which to choose. Many of the choices, however, are available on a regional basis only, because of the limited availability of the raw materials. The manufacturing process is rigidly controlled, for the most part, to provide uniformly sized brick of high strength and long durability.

The clay and water mix, along with whatever additives may be necessary for a particular batch, is poured into huge machines which compress it. The mix is extruded in a continuous bar of the proper width and thickness for the particular style of brick being made, and the individual units are cut from the bar in an unending operation. The wet bricks are stacked on pallets and moved into drying sheds, where they remain for several days under moderate heat. This results in dried bricks that are really nothing more than squared-off lumps of dirt. They have no strength whatsoever, and could easily be crumbled.

The next step moves the bricks into a kiln where they are fired at high temperature for about 24 hours. During this time molecular changes take place, the clay particles partially fuse together, and the familiar brick coloring appears. Higher temperatures and/or longer firing times mean harder bricks with deeper and darker colors. Lower temperatures and/or shorter burning times result in softer and less weatherable bricks or lighter colors such as pink and salmon. Products of the latter process are known as *underburned* brick, and are generally used more for decorative purposes than structural. They are also prone to absorb more moisture than bricks which have been fired for a longer period. Not much attention is paid to the color of the common bricks used for structural purposes, so these colors are quite variable. Where color is a consideration, time, temperature and chemical additives are manipulated to generate the desired results.

The many textures which appear on brick faces—perfectly smooth, striated, stippled, scored—are introduced after the material has been extruded, but before it has dried. In some cases, special finishes are imparted by molding the brick in forms rather than by extruding.

SIZES AND TYPES

As mentioned before, there are a great many types and styles of brick. To begin with, there are several main categories which depend primarily upon the intended use of the brick. There is some overlapping among uses, and some brick styles can be used for several purposes. The most important of the categories are *common* brick, *face* frick, and *firebrick*. A fourth group, *paving* brick or *pavers*, is used differently than the other three categories and will be considered later.

Common Brick

Common brick is just what the name implies, the most commonly used and widely available of all brick types. Common brick is made in many styles and shapes, with differently textured faces. Just about any masonry structure can be built from common brick, from walkways and barbecues to warehouses and industrial buildings. This is the least expensive brick, generally strong and durable but with specific strength, color, the textures dependent entirely upon local manufacturing conditions; there is no particular standard save for approximate dimensions and sufficient quality to be usable under normal circumstances with no problems.

Face Brick

Face brick frequently adheres better to more accurate dimensions, and is often made of higher quality clays than the common brick. Face brick is stronger and more durable than common brick, and may be treated with a variety of special finishes. These include texturing, sanding, patterning, glazing, and many more. The appearance, coloration and quality of finish are carefully controlled. The principal use of this brick is in constructing outside wall surfaces, in which instance it is often used in combination with common brick (the common brick being relegated to sections of the structure which won't be open to view or where appearance is of little consequence). Though face brick can be used in the same manner as common brick, and in fact will provide a sturdier structure, the additional cost is usually prohibitive.

Firebrick

Firebrick is seldom used for any other purpose than to line furnaces, fireplaces and chimneys, and as protection against high heat. Firebrick is extremely sturdy and durable, its great strength resulting from extended periods of high-temperature curing. Flint is added to the mix of the usual clay materials so that the resulting bricks can better withstand long periods of intense heat.

Names and Modules

As a convenience for reference and identification, the various sizes of common brick and face brick have been given names. The most common sizes are shown alongside their names in Fig. 7-1. The dimensions of all types of brick, with the exception of firebrick, are modular. Bricks, like concrete blocks, are manufactured to accord with the 4-inch module scheme, but unlike blocks, they only conform in length and width. To illustrate, the height of the brick is variously calculated so that anywhere from one to five courses are required to arrive at a dimension which accommodates to the 4-inch module principle. For instance, a 6-inch Jumbo brick is 4 inches high, so it is in itself a 4-inch module when one course high. Norwegian brick, on the other hand, is 3 1/5 inches high, so five courses are required to reach a height of 16 inches, a multiple of the 4-inch module. Actually, even the firebrick eventually conforms, if enough bricks are laid. Note that the dimensions given are nominal or trade-size dimensions. The actual dimensions are somewhat less and are predicated upon the inclusion of a 1/2-inch-thick mortar joint

Brick	Size
Standard	4 × 2 2/3 × 8
Engineer	4 × 3 1/5 × 8
Economy	4 × 4 × 8
Double	4 × 5 1/3 × 8
Roman	4 × 2 × 12
Norman	4 × 2 2/3 × 12
Norwegian	4 × 3 1/5 × 12
Utility	4 × 4 × 12
Triple	4 × 5 1/3 × 12
SCR Brick	6 × 2 2/3 × 12
6-inch Norwegian	6 × 3 1/5 × 12
6-inch Jumbo	6 × 4 × 12
8-inch Jumbo	8 × 4 × 12

Fig. 7-1. Names and sizes of popular brick styles.

to remain in harmony with the 4-inch module dimension. The module dimension, by the way, is an exact one. Also, there will be some small variation in size from brick to brick, especially with common brick, because of uneven shrinkage during the manufacturing process.

There are other less common brick styles as well, but their names are more or less regional and not widely known. Furthermore, individual manufacturers, especially some of the larger ones, often name their own products: Dutch Face, Colonial Pattern, etc. These names are narrowly localized, and have little meaning except to the manufacturer and his own customers. When selecting brick for a project, your best bet is to thoroughly investigate the local market place to see exactly to see exactly what is offered and to check upon specific dimensions. Most brick is distributed close to the point of manufacture because of weight, shipping and handling problems, so there are many variables.

BRICK BONDS

The word *bond* has a number of meanings, and there are even several differing ones as it refers to masonry. A *mortar bond* is the adhesion of the mortar within a joint, or within all the joints of a structure, to the surfaces of the masonry units and to any reinforcement materials used in the structure. A *structural bond* is the one provided by masonry units which interlock or interweave within a structure, held together both by themselves and by the mortar bond. A *pattern bond* may be structural, or purely decorative, or a combination of both. It consists of the specific pattern or patterns of masonry units and mortar joints along with their spacing, interlocking, overlapping and general positioning. Many types of pattern bonds, particularly the structural pattern bonds, have found widespread acceptance over the years and have been given their own names.

Nomenclature of a Brick

Bonding brick, that is, those bricks used in making up the various types of brick bonds, are named according to the positions they occupy in the bond. Since the following discussion needs a starting reference point, first let's look at the parts of an individual brick. Figure 7-2 shows a typical brick in what might be called a "normal" position. This brick has cores, but many bricks do not. Cores do not diminish the strength of a brick, but do make it a bit lighter and thus easier to handle. Some brick have slight depressions instead of cores; this feature helps improve the mortar bond. The two surfaces through which the cores run are the top and bottom of the brick. If the brick is a bit wider on one of these surfaces, the wider surface is the top, and the narrower one, the bottom. The

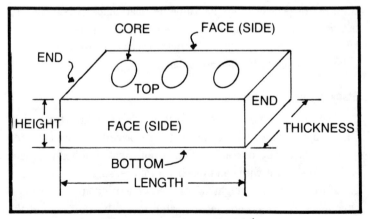

Fig. 7-2. Nomenclature of a brick.

dimension between top and bottom is the height. The long and narrow surface facing you is a face, and the opposite is also a face. If one face has a special finish, such as stippling or glazing, this is called the finished face and in structures is the face always presented to view. The dimension between the two faces is the thickness of the brick. The remaining two surfaces are the ends, and the dimension between them is the length of the brick.

Positional Names

When a brick is laid in a bond in the position shown in Fig. 7-2, it is called a *stretcher*. In most bonds, stretcher bricks make up the largest proportion of the total. If the brick is installed bottom-down and end-to, it is called a *header*. If the brick is stood up on end and presents a vertical face, it is called a *soldier* brick. Bricks may also be laid with one face down and the other up. When the bottom or the top of the brick is exposed in the bond, the brick is called a *shiner*. If an end is presented, the brick is called a *rowlock*. A brick standing on end and presenting either top or bottom is a *sailor* brick. These relative positions are illustrated in Fig. 7-3.

A fractional length of brick, resulting from the trimming of the unit to fit into a particular bonding pattern, is designated by the size of the piece to be used relative to a full brick, and is called a *bat*. Thus, three-quarters of a brick is a 3/4-bat and the piece remaining is a 1/4-bat. Depending upon the circumstances of construction, partial bricks may also carry a name based on use and position. *Closure* bricks, which often assume peculiar shapes when finishing out corners or odd designs, have their own special names, such as *king closure* or *queen closure*. Bricks which are used in corners, either fractional or whole, are *quoins*.

Stacking and Overlapping

There are two general ways to assemble a structure from brick. One is by *stacking*, where the units are placed one directly atop another. This type of construction is not capable of bearing much weight, nor will it withstand much lateral pressure, and so is relegated to use in lightweight decorative construction and veneers, although it is sometimes incorporated into other portions of the projects where structural strength is already provided. The other and by far the most common method is the *overlapping* assembly. With this construction, each masonry unit both overlaps and is overlapped by another unit. The greater the overlap, the greater the bond and the consequent strength of the assembly. Reducing the overlap to zero results in a stacked bond of the lowest possible strength, but which, however, can be beefed up by the inclusion of metal ties.

Overlapping is usually designated in terms of the fraction of the unit overlap of the stretcher bricks. Thus, where half a brick overlaps half of another brick on the course below, the result is a 1/2-lap

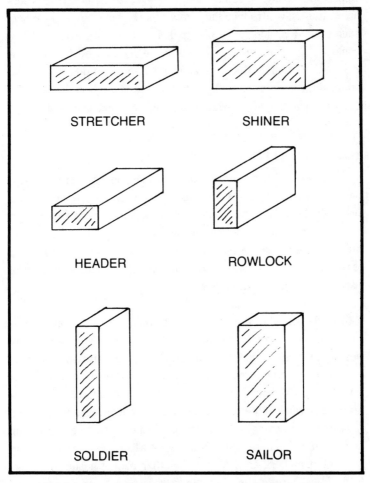

Fig. 7-3. Names of bricks depending upon their laid-up position.

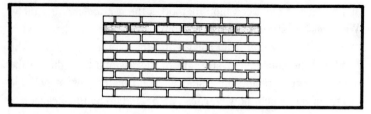

Fig. 7-4. 1/2-overlap bond. Courtesy of the Brick Institute of America.

method of laying (Fig. 7-4). The most common methods are 1/4-lap, 1/2-lap, and 1/3-lap, though there are plenty of variations. The lapping may change schemes along a single course, as when headers are periodically introduced in a row of stretchers; an arrangement such as a 1/4-lap and then a 3/4-lap, followed by a return to a 1/4-lap, could result. Nor need each course be overlapped in the same manner. Changing the overlapping system periodically as the courses rise also changes the pattern bond of the wall. This contributes a decorative effect which can be further enhanced by the use of two or three colors of brick.

Traditional Bond Patterns

Numerous bond patterns have become popular over the years, and are now considered standards. In some cases these bonds are primarily decorative, but in most constructions they are assembled as structural pattern bonds and are responsible in part for the strength and rigidity of the structure. There are also a great many lesser-known and unnamed bond patterns and a wide assortment of patterns with minor variations.

One of the most popular bond patterns is the *running bond* (Fig. 7-5), widely used in the brick-veneer construction of homes. The standard variety consists entirely of 1/2-lap stretcher courses; no headers are used. A common variation is to lay the brick in a 1/3-lap.

Another common pattern is the *American bond* (Fig. 7-6). Used structurally, this pattern provides a strong construction and a pleasing pattern while at the same time requiring only a minimum number of brick. In its strongest form, the pattern consists of four rows or courses of stretchers and a fifth row of all headers. Another four rows of stretchers are laid, then another row of headers, and so forth. If less strength is allowable in the structure, the header row

250

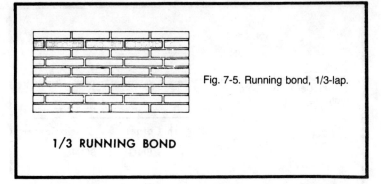

Fig. 7-5. Running bond, 1/3-lap.

1/3 RUNNING BOND

can be moved up to the sixth course, or even the seventh. Each row of headers must be balanced with a pair of 3/4-bats to conform to the 4-inch modular spacing and so that the row will come out even. These bats are placed one at each end of the course.

The *English bond* (Fig. 7-7) consists of alternating courses of stretchers and headers, and is a strong construction. A header is centered upon the stretcher below, and the two adjacent headers are centered over the vertical stretcher joints. The result is that all of the vertical joints of the stretchers line up. Uniformly sized brick must be used, along with a considerable amount of care in mortaring, so that the joint lines will be accurately positioned.

The *Dutch bond* (Fig. 7-8) is a variation of the English bond, and in fact is sometimes called the *English cross bond*. Stretcher courses are alternated with header courses, but the difference lies in the positioning of the headers. The vertical stretcher joints do not line up from course to course, but are staggered so that only every fourth course aligns. This results in the characteristic cross pattern, which can be greatly emphasized to make a decorative effect by using a

Fig. 7-6. American or common bond with sixth-course headers.

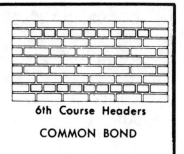

6th Course Headers

COMMON BOND

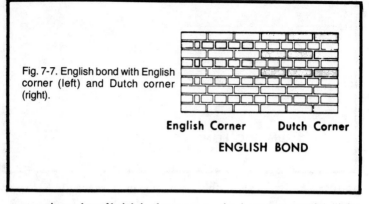

Fig. 7-7. English bond with English corner (left) and Dutch corner (right).

English Corner   Dutch Corner

ENGLISH BOND

contrasting color of brick in the crosses. Again, accuracy of the joint lines is important in making a uniform pattern.

The *Flemish bond* (Fig. 7-9) is another popular pattern into which a good many variations can be introduced for decorative purposes. This also is an extremely strong construction. Each course consists of alternating stretchers and headers. The arrangement is shifted in successive courses so that each header is centered upon the stretcher above and the stretcher below. This results in the appearance of interlocked stacks. The same pattern can be used to make a decorative screen wall by eliminating the headers and leaving gaps where the headers would normally be placed. Such a construction is inherently weak, however, and should be done in sections or panels supported by solid masonry columns.

The *stacked bond* (Fig. 7-10) is an inherently weak construction primarily used for decorative or architectural effects, though it can be beefed up somewhat by the use of reinforcement. In this bond the

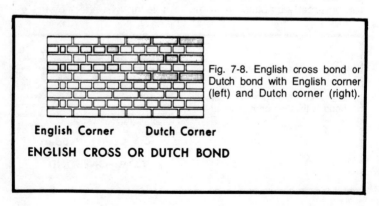

Fig. 7-8. English cross bond or Dutch bond with English corner (left) and Dutch corner (right).

English Corner   Dutch Corner

ENGLISH CROSS OR DUTCH BOND

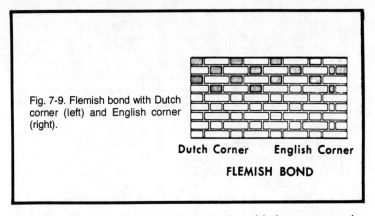

Fig. 7-9. Flemish bond with Dutch corner (left) and English corner (right).

Dutch Corner   English Corner

FLEMISH BOND

joints line up both vertically and horizontally, with the masonry units stacked one above the other in straight rows. The brick can be stacked in any of the six positions; stretcher, header, soldier, shiner, rowlock or sailor. Any or all combinations of positions can be worked out for different patterning effects.

Pattern Bonds for a Garden Wall

One of the best constructions for the home mechanic who wants to build a sturdy wall is a variation on the Flemish pattern called the *garden wall bond*. Assembly is easy, and the pattern comes out the same on each side of the wall. There are three common types of garden wall bond; 2-stretcher, 3-stretcher, and 4-stretcher. This is a double-thickness structure which interlocks nicely for a great deal of strength and stability, important factors in freestanding walls.

A 2-stretcher garden wall is made by laying a pair of stretchers end to end, followed by a header, followed by another pair of

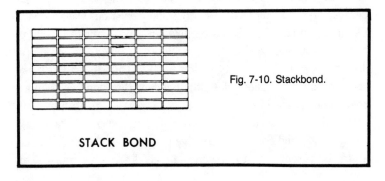

Fig. 7-10. Stackbond.

STACK BOND

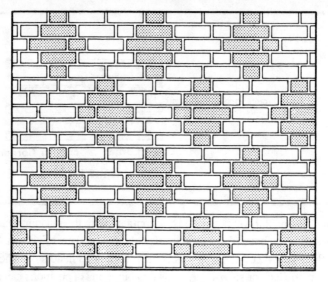

Fig. 7-11. Two-stretcher garden wall bond.

stretchers, and so forth, following the same pattern in each course (Fig. 7-11). The second course is shifted so that the headers are centered over the stretcher joints in the first course.

The 3-stretcher and 4-stretcher walls are made in much the same way, using three stretchers or four stretchers in a row, respectively. Liberal use of bats must be made to make the courses come out even and to maintain the proper joint alignment. Various patterns can be worked out in the process, and emphasized with contrasting colors of brick if desired.

Variations on the Standard Bonds

The bonds described thus far are the traditional ones. But because of the modular nature of brick and the various ways in which the units can be placed in a structure, an almost infinite variety of patterns is possible while at the same time a high degree of structural strength is maintained. Imagination and a willingness to do a fussy and careful job are about all that is needed. By using two or more colors of brick and changing their positions within the struture, you can work up all sorts of geometric designs, such as diamonds, triangles, X's and crosses, repetitive patterns and the like. Diagonals can be introduced into a common bond (all stretcher courses) by alternating light and dark stretchers in each course. The diagonals themselves can be altered by changing the lap pattern. Other patterns are created by using light-colored stretchers and dark headers or vice versa. Contrasting headers might be inserted at only certain points, making a wide-spread pattern. Diamonds can be made with a dark center, medium-tone middleground and dark outline, all set in a light-colored matrix.

Choosing Your Bond Pattern

Exactly how the bricks are to be positioned, the various colors and shadings are to be used, the patterns themselves and how they are to be related to the remainder of the structure, symmetry and repetition of the pattern, etc., are decisions which will be made entirely at the discretion of the designer. Skill and taste are needed to put such intricate designs together, but the task is by no means beyond the ability of the do-it-yourselfer. Once the design is formulated, the execution is a matter of careful and precise bricklaying and good workmanship.

You can make your own designs readily enough, starting with a design drawn on paper. First decide what kind of brick you are going to use, and determine the exact dimensions, including mortar joints. Decide upon the overall size of the wall or structure that you want to build, and draw its outline to scale on a sheet of paper. Scale the brick units to the same size and draw in the courses as you envision them, working out your patterns as you go along. The process is rather like making a mosaic, or perhaps drawing a maze. Keeping an accurate scale will allow you to maintain the proper dimensions and proportions; you'll even be able to determine the sizes of any necessary bats. If your finished design uses but a few headers, you can insert metal ties or metal reinforcement in their stead for additional strength.

Another way to make up your drawing, if you already have a preconceived central design in mind, is to first lay out the brick units which will compose the design itself, and then fill in the remainder of the wall with whatever subdesigns may seem appropriate. Bear in mind, too, that not only the joint pattern but the mortar joints themselves can also become a part of that design by virtue of using one or several colors of mortar. Texturing, glazing or other brick face finishes can also play a part.

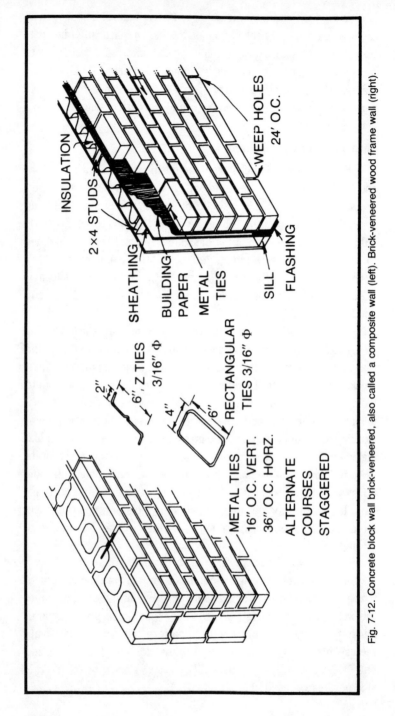

INSULATION

2×4 STUDS

SHEATHING

BUILDING
PAPER

METAL
TIES

SILL

FLASHING

WEEP HOLES
24' O.C.

2"

6" Z TIES
3/16" Φ

4"

6"

RECTANGULAR
TIES 3/16" Φ

METAL TIES
16" O.C. VERT.
36" O.C. HORZ.

ALTERNATE
COURSES
STAGGERED

Fig. 7-12. Concrete block wall brick-veneered, also called a composite wall (left). Brick-veneered wood frame wall (right).

BUILDING WITH BRICK

There is a vast array of structures that can be made from brick, since it is the most versatile and flexible of all of the masonry unit building materials. There are a number of projects which lend themselves nicely to the dry-laying process, where the bricks are stacked and interlocked in an appropriate design without benefit of mortar. In the interest of stability and durability, though, most brickwork is laid up with mortared joints. For all such construction, footings are necessary. Brickwork generally takes the form of a wall, or some modification thereof. There are three general methods for building brick walls.

Brick Veneer

A *veneer wall* is constructed as a facing for a substructure, which may be of wood frame or masonry construction (Fig. 7-12). The brick is usually laid longitudinally, presenting the narrow face to the outside. Unless one of the special types of brick is used, the course is only 4 inches thick. In all cases, the veneer is attached to the substructure with corrugated metal ties or some similar hardware. One end of the tie is nailed to the substructure, while the corrugated portion is bent outward and embedded in the mortar joints. One tie is used for every 2 square feet of wall area. Usually a space of about 1 inch is left between the interior of the brick veneer and the exterior of the substructure. This space is most often left empty, but in some types of construction is filled with grout or mortar.

Cavity Walls

Hole or *cavity walls* (Fig. 7-13) are made up of two separate sets of courses laid in a running bond of stretchers only, with bats sometimes added to provide a bonding pattern. There are no interlocking masonry units between the two separate sections, and the space between the two varies from 2 to 3 inches. Both sections are laid upon a common base or footing. For greater strength and rigidity, the two sections are linked together with special metal wall ties made for the purpose.

Cavity walls are generally intended for structural purposes rather than decorative. Their structure was designed to provide two

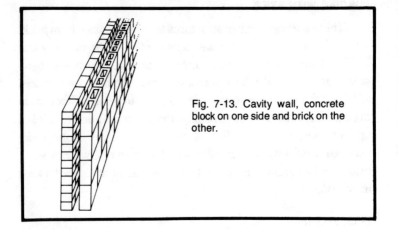

Fig. 7-13. Cavity wall, concrete block on one side and brick on the other.

important and special features: moisture resistance, and thermal insulation properties. In areas of high rainfall, any moisture which seeps into the outer wall section appears on the inner surface of that section, rolls down the wall and drains back to the outside through weep holes provided along the base of the wall. Moisture cannot penetrate through the inner wall, so the interior of the structure remains dry. Thermal insulation is provided by the air space between the two wall sections. In some cases, this space can be filled, or partly filled, with thermal insulating material for improved performance. This must be done in such a way that the material does not interfere with moisture drainage. A cavity wall can also be used to build decorative or nonstructural walls where added thickness without the additional expense of extra brick is desired. Though standard cavity wall tie wires are designed for a 2- to 3-inch cavity (this being the optimum air space), special tie wires or reinforcing arrangements can be worked out so that the two sections can be set farther apart. The top of the wall is then capped with solid masonry units.

Solid Walls

Solid wall construction is most commonly used in such home projects as garden walls, patio seating walls, barbecues and any other kind of project where solid construction is the easiest course. There are instances where a 4-inch wall is adequate, and perhaps desirable. This thickness is not strong, however, is not especially rigid, and is susceptible to mechanical damage, especially from lateral pressures. This type of construction often demands that the 4-inch wall sections be further supported by columns or set as panels bonded by structural masonry much in the manner of framing a picture.

The 8-inch wall is widely used in all sorts of wall construction, and has sufficient strength and rigidity to be adequate for virtually all purposes around the home. Full-length headers or rowlocks are easily interspersed with stretchers and shiners to make a fully interlocking assembly. Such a wall can be constructed in any of a number of pattern and structural bonds.

Brick walls can be assembled in any thickness conforming to the 4-inch module principle. The 12-inch wall is common, and so is the 16-inch and 20-inch. In some industrial constructions of earlier days, when both brick and labor were far less expensive than they are now, walls 6 feet thick or more were not uncommon.

CORNERS

Forming the end of a brick wall is a relatively simple process, and, in most cases, if the bond is properly laid out and the first course established correctly, the end will square off in natural fashion. Building corners, however, is probably the trickiest part of making brick walls. Though many corners are right-angled, either obtuse or acute angles can also be constructed. Right-angle corners are generally a bit simpler, but a certain amount of cutting is frequently necessary in any corner.

Starting the Corner

As in concrete block construction, the corners (or ends) of the wall are the starting points. The first few courses of the corner, depending upon the bond pattern and until the pattern begins to repeat, are the starting point of the bond and determine whether or not, and how well, the pattern will work out. Exactly how the corner is started depends upon the structural bond and/or the pattern bond that appears on the wall face, as well as the thickness of the wall.

There are a great many variations of the corner-starting procedure, far too many to treat in detail here. One good rule to follow, for all corners, is to make sure that the brick which forms the outermost corner itself is never smaller than 4 inches in width, and is preferably

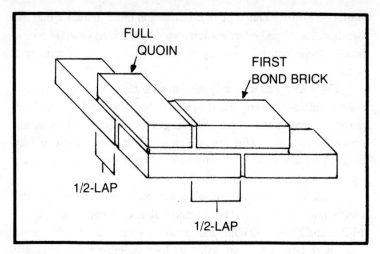

Fig. 7-14. Corner construction for 1/2-lap bond.

a 3/4-brick or quoin (and wherever possible, a full brick or quoin). Bats of appropriate size can be used adjacent to the quoins on the outer surface, or adjacent to full bricks on the inner corners, to act as closers between the quoins and the beginning of the bond bricks proper. In other words, set the corners with bricks as near full-size as possible; leave whatever gap is necessary at either the inside or the outside face, or both; start the bond pattern off with full bricks; and fill in the space between with a chunk of brick of whatever size is necessary. Figure 7-14 shows a simple corner construction for a 1/2-lap bond where all the bricks remain full-size. Figure 7-15 shows a somewhat more complex corner for 3/4-lap bonding, where some cutting is required.

Angle Corners

Obtuse and acute angle corners present their own special problems. In some areas you can purchase brick which is specially formed for 30-, 45-, and 60-degree angles, and this makes the task simpler. For all other angles, however, the brick must be cut and fitted as necessity dictates. In all corners of this sort, the object is to cut the brick as large as possible and use a minimum number of small closers and bats, while at the same time not disrupting the bond pattern any more than is necessary. Obviously the corners will present a different pattern than the face bond, but if the joints are kept straight and uniform, the pattern maintained throughout the corner section, and opposite corners or ends made identical, the whole effect will not be unpleasing to the eye.

There are three possibilities for forming acute and obtuse angle corners. One is to cut all of the bricks at all inner and outer angle points so that a clean and smooth face is presented on both sides of the wall. This is the recommended practice and should be used wherever possible, especially if the wall happens to be a load-bearing one. This method insures that the angle corner will be weatherable and very durable.

Another possibility is to stack the bricks so that successive courses overlap one another and the uncut ends of the bricks stick out on the outside face of the angle, with alternating successive bricks cut to a matching pass angle on the inside face. This results in a vertical series of alternate projecting brick corners facing in each of the two angle directions. The third possibility is to stack the bricks so that their corner edges meet at the angle. Each alternating outside corner in the successive courses lines up, and there is an alternating series of triangular pockets in each wall face. Both of these methods are used in making obtuse angles and, in both cases, the inside face of the angle is smooth with no projections *or* pockets. A style which incorporates both projections *and* pockets is the "pigeonhole" corner used to make acute angle corners.

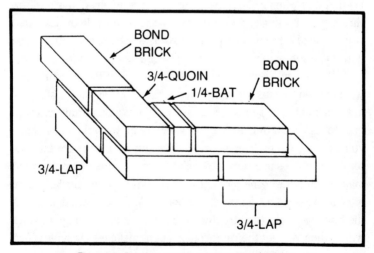

Fig. 7-15. Corner construction for 3/4-lap bond.

All of these methods are great time savers, but do not result in structurally sound corners. On the other hand, they do have a certain amount of decorative merit, and at the same time are much simpler for the do-it-yourselfer to build than fully-cut corner constructions. There seems little real reason not to use any of these corner constructions, especially for nonload-bearing walls built in areas where the weather conditions are benign, provided that good workmanship is used and the mortar joints are well bonded and tooled to minimize weathering.

Planning for Bonds and Corners

The pragmatic approach is about the best one for the home mechanic to take when trying to sort out the sometimes confusing problems connected with the stacking up of bonds and corners so that everything is even and presentable. There are two good methods and, if you wish, you can combine them. One is to draw the whole business out to scale, using your choice of a structural or pattern bond and fitting the corners together in one or another of the "standard" corner or end constructions. Remember you must contend with three surfaces: the viewed face, the reverse face (which may or may not be visible), and the corner or end itself. In some situations there will actually only be two surfaces to worry about, but in others, as when a blunt-end obtuse angle is involved, there will be three. If everything is drawn to exact scale, you will know just what you will need in the way of quoins, closers and bats, and you can also determine the number of bricks you will need per course.

The second method is to chalk the outline of the structure you want to build on your patio, driveway or some other flat surface. Lay in a stock of brick, and stack them up without any mortar, and with gaps to represent the vertical joints. By trying various combinations you can work out both structural and pattern bonds that are satisfactory, and at the same time determine just what cuts you will have to make to form the corners or ends. This system is a particularly good one for laying out intricate pattern bonds. As long as you don't stack the brick too high, the structure will remain stable enough to work with. When you get the right combination, make a sketch for referring to during the actual construction.

OPENINGS

Openings in brick walls are handled in much the same way as those in concrete block walls. It is by far preferable to position and size the opening to coincide with the vertical joints of the bond. This makes for a stronger structure, and requires less labor and fussing around in the construction. Since it is highly important that a footing extend uninterruptedly under each section of a discontinuous wall, wall openings will receive a footing incidentally, even though they don't actually need one. The uninterrupted footing affords greater stability for the wall and provides a base for a sill to be installed in the opening if desired. In any event, if the openings are closed across the top or bottom, or both, the footing must be continuous, anyway. The sides of open-topped openings are treated as though they were ordinary wall ends.

Lintels

Where the wall openings will be installed with some sort of closure, such as a door or window, the treatment is again much like that for concrete blocks. Lintels must be installed above window, viewport and door openings. Heavy wood beams which extend a minimum of 4 inches into the wall on each side are sometimes used if the opening is not more than 3 feet wide. Concrete lintels may be used, but only if they are of the precast reinforced variety. Steel, either in the form of angle iron or specially formed shapes, is perhaps the most common choice. Special reinforced brick can also be used in some circumstances.

Arches

Another possibility, one which should not be tried by the amateur mason until he gains considerable experience with brickwork and mortar handling, is the brick arch. Arches can be made circular or in a variety of ellipses, and if properly made they are tremendously strong and will bear a great deal of weight. Their construction, however, is not simple. Spanish arch openings as viewports in garden or patio walls are handsome features.

Sills

Window sills can be made of concrete, but the results aren't satisfactory unless the sill is precast and reinforced. Stone, particularly granite, was often used years ago, but seldom is today. The most common practice is to use the same brick that makes up the wall itself. In an 8-inch wall, for instance, a row of brick is set in shiner position flush with the inside face, while a second row is set in rowlock position to the outside face, tilted downward at a 9 1/2-degree angle and protruding slightly past the wall face. In other types of installations, specially-formed brick of various dimensions is available for making sills.

Door sills are usually made of precast and reinforced concrete, set in place so that they extend into the masonry walls on each side. In many installations, the sill is omitted, and flooring or paving is carried straight through the opening. Small thresholds of marble, granite, wood or some other building material may be installed if desirable.

Framed Openings

Window, door or viewport frames can be installed by setting them into the openings after the mortar joints in the wall structure have cured. They may also be set in place and braced as the construction proceeds, and the brick built up around them. The frames themselves are generally made of wood, aluminum or steel. Special types are available which fit into the brickwork and are sized in both height and width to conform to brick dimensions and to minimize the amount of cutting.

ESTIMATING NEEDS

There are a number of ways of estimating the number of brick you will need for your project. If you have designed the bond to scale first, then you can actually count the number of bricks needed, adjusting as necessary for a 4-inch, 8-inch or 12-inch wall. If the project is a fairly large one, determine the area of the walls and then consult the table in Fig. 7-16. This table assumes that all the brick are laid in stretcher position, so you must make adjustments if any of the brick will be laid in different positions. For instance, if every other course in an 8-inch wall is a header course, you would need 50 percent more brick than if they were all stretchers.

Another method of figuring is to determine the number of brick needed in each course of a particular bond pattern and then multiply by the number of times the pattern repeats itself. Or, find the area of one brick, including mortar joints. Calculate the total area of the wall or other surface in square inches. Divide the total area by the brick area and then multiply by the number of courses needed to make up the full height of the structure. Then multiply this answer by the number of layers deep the wall will be; by one for a 4-inch wall, by two for an 8-inch wall and by three for a 12-inch wall. Again, these figures must be adjusted to compensate for bricks which are laid in positions other than as stretchers.

Whatever method you use to figure the number of brick needed, be sure to allow for an extra 5 percent when you order. This will cover those that are cracked or damaged, and also will replace the ones which are improperly cut or broken during the construction process.

By referring to the table in Fig. 7-16, you can also estimate your mortar needs. The figures in the table take into account the fact that

	4-inch Wall		8-Inch Wall	
Area of Wall	Number of Bricks Required	Mortar, Cu. Ft.	Number of Bricks Required	Mortar, Cu. Ft.
1	6.2	0.075	12.4	0.195
10	62	1	124	2
20	124	2	247	4
30	185	2.5	370	6
40	247	3.5	493	8
50	309	4	617	10
60	370	5	740	12
70	432	5.5	863	14
80	493	6.5	986	16
90	555	7	1109	18
100	617	8	1233	20
200	1233	15	2465	39
300	1849	23	3697	59
400	2465	30	4929	78
500	3081	38	6161	98
600	3697	46	7393	117
700	4313	53	8625	137
800	4929	61	9857	156
900	5545	68	11089	175
000	6161	76	12321	195

Fig. 7-16. Brick and mortar estimating table.

nominal brick sizes include an allowance for a 1/2-inch mortar joint thickness in accordance with the 4-inch module scheme. Actual brick dimensions, however, vary slightly from maker to maker, so the mortar joint may have to be a bit thicker or thinner to allow for the proper dimensions. If you propose to use 3/8-inch joints, the figures in the table should be reduced by 25 percent. If you go to a 5/8-inch joint, increase the figures by 25 percent. Further adjustments may have to be made if you use one course of face brick backed by one or more courses of common brick, thereby requiring mortar joints of somewhat different thicknesses. To determine the particular mortar mix and the quantity of materials necessary to make that mix, refer to the mortar section in Chapter 5.

LAYING BRICK

Brick is laid up in much the same fashion as concrete block, with only a few minor differences. Those differences, however, are important ones. The preliminary work is identical. After the project is designed and checked over, acquire all the necessary materials. Prepare the site, and build any necessary footings. After the footings have cured, lay out the tools and equipment and stack the brick at convenient locations on the job site.

The first step in the brick constructions is to clean off the footing tops and locate the position of the first course. The centerline of the brick structure should fall on the centerline of the footing for walls, and be appropriately laid out for other structures. A chalkline is handy for the layout work. Set out the first course of brick dry, leaving spaces to represent the vertical mortar joints. Check to make sure the dimensions and the bonding pattern are correct.

Wet or Dry Bricks?

The next step involves an aspect of bricklaying which is markedly different than block laying. Concrete block is never under any circumstances moistened before laying, because the block swells and will later shrink and destroy the joint bond. With brick, however, the situation is frequently just the opposite. But not always, and brick should always be checked first to see whether or not it should be dampened. The problem is that some brick is too absorbent—it depends upon the length of time the raw materials were fired in the kiln. Strongly absorbent brick will suck water out of the mortar and cause a weak mortar joint. Hard brick are only slightly absorbent and can be laid dry, but the others must be moistened first.

The Brick Institute of America recommends the following procedure for checking the water absorption of brick. Select a brick at random from the pile and draw a circle about the size of a quarter on one face. Drop 20 drops of water from a medicine dropper inside the circle. Wait 1 1/2 minutes. If you can still see water on the brick surface at the end of that time, lay the brick dry. (Also, cover your stockpile so the brick will remain dry in case of bad weather.) If the water has disappeared into the brick, the brick should be laid while damp, but not dripping wet. Soak the pile of brick with a garden hose about 15 minutes before you start using them, and repeat the process whenever necesssary as the job progresses.

Mortaring Procedures

With everything in readiness, mixing the mortar is the next chore. The various mortar mixes and procedures are the same as for concrete block work, and further information can be found in Chapters 5 and 6.

To start the first course, throw a line of mortar on the footing top sufficient to take care of the first three to five bricks. If the course is composed of stretchers, the mortar will be thrown in a long line. Spread the mortar evenly by furrowing with the trowel tip, just as for concrete block. Do each wall thickness separately; the outer face first, the middle (or the inner face for an 8-inch wall) next and then the inner face. Alternate back and forth as necessary. If the course is all headers, the first mortar bed will be a rectangular patch. For alternating stretchers and headers, the same is true. Make sure that the mortar bed is thick enough so that when you press the bricks into place the joint will be 1/2 inch thick. This is a full mortar bed, and should completely cover the bottom of each brick (Fig. 7-17).

Setting the Bricks

The proper way to grasp a brick is to hold it between the thumb on one side and the fingers on the other, with the brick horizontal and the bottom facing downward. Most right-handed masons will set brick with their left hand, leaving the right hand free to manipulate

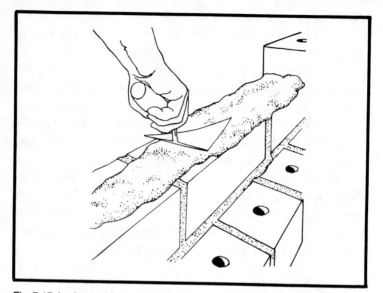

Fig. 7-17. Laying and furrowing a full mortar bed on brick. Courtesy of the Brick Institute of America.

the trowel. Pick up the first end brick and lower it directly into position in the mortar, squeegeeing it just slightly as you make contact so that the brick is set firmly in place. Make a quick visual check to see that the brick is approximately level, plumb and properly aligned and immediately set the next bricks.

Pick each brick off the pile and butter one end (or side, depending upon the position of the brick in the course). The entire end or side should be covered with a liberal coating of mortar, more than enough to make the 1/2-inch joint. With a combined forward and downward motion, shove each brick into place against its neighbor with a smooth motion so that the mortar joint is made simultaneously on the bottom and at the end or side. Urge the brick downward and inward and let the mortar curl out of the joints until the proper thickness is reached. This technique of placing brick is called the *shoved joint* method (Fig. 7-18).

Another method, generally more difficult for beginners, is the *pick and dip* method. The mason picks a brick from a pile and at the same time loads his trowel with mortar. In one continuous motion, he throws the mortar bed for a single brick and uses the remaining mortar to butter the end of the brick he is holding. He then slips the

268

brick into position, immediately strikes off the excess motar from the joints and picks up the next brick.

A third method is known as *slushing*, a practice which is common enough but not recommended. It consists of depositing a gob of mortar atop the juncture of two bricks and squeezing it downward until it fills up the open head joint. This is a sloppy way to make joints; furthermore, it is structurally unsound. It also consumes a lot of extra mortar.

As soon as the first three to five bricks are set into the mortar bed, quickly check them for level, plumb and alignment in the same fashion as for concrete block. Tap the bricks into position gently with the trowel handle, but take care not to jerk them around in the mortar bed lest the joint bond break. Take plenty of time and care with this first course, because as it goes, so goes the rest of the construction. If you are not satisfied with the way things turn out or can't achieve proper alignment, take up the bricks, remove the mortar and start over. The extra time spent is well worth while. As soon as you are satisfied, go on to lay the next three to five bricks in the same course.

As with concrete block, the corners or ends are the starting points and are built up first, and the space between filled in after-

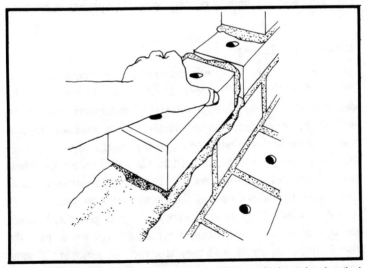

Fig. 7-18. Shoving brick into place to make simultaneous horizontal and vertical mortar joints.

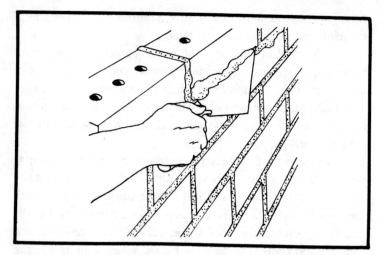

Fig. 7-19. Striking excess mortar from the joints.

ward. As soon as a few bricks in the first row are laid, go on to the second course. The second course, and all subsequent ones, is laid in exactly the same manner as the first row. Use a full mortar bed and full head joints, striking off excess mortar as you go along (Fig. 7-19). Follow your structural or pattern bond design, cutting the corner brick as necessary. Take plenty of time to align, level and plumb the bricks; plumb all corners regularly with a straightedge.

Alignment and Tooling

After the corners or ends have risen a few courses, you can begin to fill in the space between. Set a mason's line as a guide for each course. The line should be offset 1/32 inch from the face of the wall. Each brick should be aligned parallel with the mason's line, with the top of the line even with the top of the brick. Keep the mason's line taut so there is no sag in the middle. Closure brick are installed in the same manner as closure block, by buttering the bottom and sides of the opening and the ends of the closure brick as well.

As sections of the structure are completed, periodically check the mortar joints to see if they are hard enough yet to pass the thumbprint test. As soon as they are, tool the joints in the same manner as you tool concrete block joints (Fig. 7-20). The implements used are the same, and so are the various joint styles.

270

ANCHORING

Anchors in brick work are used for two principal purposes. One is to reinforce and tie together adjoining brick structures, such as intersecting walls and wall corners. The other purpose is to tie a superstructure to a brick base, such as a roof plate atop a carport wall. Anchoring adjoining brick structures is often done with #3 reinforcing bar, the same as is used for reinforcing poured concrete. The bar is laid into the mortar joints in appropriate positions, extending a considerable distance into both structural sections. Other types of anchors, both functional and decorative, are made for insertion into one structure as it is coursed up, with the stub end being embedded in the joining structure during its construction.

Where a plate or other superstructure is attached to a wall top, steel anchor bolts of a minimum 1/2-inch diameter and 12-inch length are embedded in the mortar joints, with the threaded ends protruding just enough to clear the plate and accept a washer and nut. These anchor bolts should be placed every 4 to 6 feet along the length of the plate. Smaller and lighter superstructures can be attached with bolts arranged to come up through the mortar joints or through mortar-filled cores in the brick.

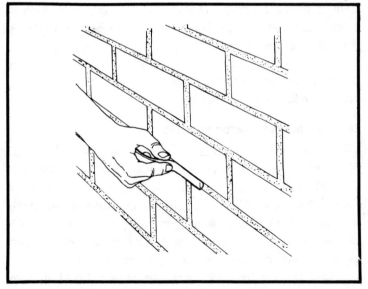

Fig. 7-20. Tooling mortar joints in brick wall.

All bolts should be bedded as deeply as possible into the wall or other structure, so as to afford the greatest amount of anchorage. A common mistake is to set anchor bolts so that all of the stress and strain is transmitted to one or two bricks in the top course. The inevitable result is that both anchors and bricks work free if a little too much pressure happens to be applied to the superstructure. Make sure that all ornamental iron work, grilles and screens, lamp posts and similar items are solidly secured, and their anchors surrounded by as much brick and mortar as you can manage.

CLEANING

When putting together a brick structure, cleanliness is an important factor. Most brick, especially those with a rough surface texture, have a propensity for picking up spots and stains which are difficult to erase later. One of the reasons for using brick in the first place, especially for small projects around the home, is for their aesthetic value. It makes sense, then, to keep the appearance of the brick as presentable as you can.

Dirt and Stains

When you stockpile brick prior to beginning a project, stack them up off the ground on boards or pallets, and cover them to prevent dust and dirt from falling or blowing onto them. As you lay up the brick, check each one to make sure that it is free from dust and particularly mud, and check the face that will be exposed for defects and stains. Not only is the dirt unsightly; it will also interfere with proper bonding of the mortar joints.

Removing Unwanted Mortar Remnants

As the construction proceeds, strike excess mortar from the joints carefully so that the mortar does not drop away in a blob onto the brick faces a few courses below. It is a good idea to spread a piece of plastic or an old tarp along the bottom course to prevent the brick from being splattered with mortar droppings—a few gobs will inevitably get away from you. As you tool the joints, work with care and brush away mortar droppings and crumbs as you go.

After the job is complete and the mortar joints have had ample time to cure, wash down the structure with a strong stream from a garden hose to remove loose crumbs of mortar and cement dust. When the water dries, you may find that despite all your precautions there are some mortar stains here and there. These can be removed with an application of hydrochloric acid. The solution is made by mixing 1 part commercial muriatic acid with 9 parts water, always pouring the acid into the water and never the reverse. This solution is nasty to work with, and safety precautions are absolutely essential. Always wear long trousers, long-sleeved shirt with sleeves rolled down, gauntlet-type gloves, and a face shield or at the very least, safety goggles. Muriatic acid is bad enough on the skin, but is a disaster in the eyes.

Start the process by soaking the structure thoroughly with a hose. Then swab the acid onto a 15- or 20-square-foot area with a long-handled stiff-bristled brush, scrubbing the solution in thoroughly. Work rapidly, and immediately, as the small section is done, wash it down with a strong stream of water from a hose. Continue the washing until you are sure that the area has been totally flooded with water and all remnants of the acid have been flushed away. If the acid remains in place too long, it will attack the mortar joints. All wood and metal parts of the structure must be protected from the acid by masking off with tape and construction plastic.

Efflorescence

There is a peculiar condition called *efflorescence* which sometimes appears shortly after the brick structure has been completed. This is a crystalline deposit, usually white and sometimes fluffy and also harmless, which appears in patches on the masonry surface. Some people consider the effect not at all displeasing, though others find it most unattractive.

Efflorescence is caused by a leaching out of soluble salts that are contained in the masonry materials. To reduce the possibilities of efflorescence, you can take a few precautions during the construction of the brickwork. Build with masonry units which are known to not effloresce, and mix the mortar with well-washed sand and with water which is free from any kind of mineral salts. Note that in many areas even drinking water is sufficiently loaded with chemical salts to be a potential cause of efflorescence. Make sure, too, that all mixing tools are clean and that the lime you use for mortar mixing is a hydrated lime. Mortar joints should be well tooled, preferably in a vee or concave shape.

If efflorescence does occur, it can often be removed by scrubbing the patches vigorously with a stiff-bristled brush and then flushing the wall with water. If the condition recurs, it will be less severe and the powder can be brushed away again, until eventually the efflorescing will stop completely.

If brushing doesn't do the job, the only other recourse is to treat the surface with muriatic acid. Use the same solution and the same precautions as for cleaning new brick, unless the mortar joints are colored. In that case, use a much weaker solution. Soak the surface with water first, then scrub the muriatic acid on with a stiff-bristled brush in small areas not exceeding 3 to 4 square feet. Let the solution stay for about 5 minutes, then scrub the area thoroughly with a brush. Flood the surface with water to flush away the acid and the salts. Whenever acid is used to clean a masonry surface, the entire surface should be done and not just the stained or effloresced areas. If only small patches are treated, the result may be a mottled effect which is nearly impossible to correct.

Miscellaneous Masonry Units

Though brick and concrete block are the most widely used of the masonry units—at least in sheer numbers—there are several other types that are also popular. Some of these units can be employed in much the same way as block or brick, while others have their own particular uses. Most of the same basic skills and procedures are used in assembling these miscellaneous masonry units, but a few special techniques and design parameters are also entailed. The differences will be pointed out during the discussion of these types.

PAVERS

Pavers is a general term given to masonry units which are primarily intended for creating patios, terraces, walkways, driveways and any other similar flat or slab-like surfaces. Most of these units are made especially for the purpose, and are either shaped so that they naturally interlock or are laid in interlocking or patterned fashion. Pavers are particularly easy to work with, ideal for certain kinds of do-it-yourself projects around the home, and provide a considerably different visual impact than the other masonry units. There are two types of pavers readily available: brick and concrete.

Brick Pavers

There are many kinds of brick pavers which are made particularly for paving, as regards both size and specifications. There are also some kinds of face brick, the same as are used in other types of construction projects, which can be utilized for paving purposes. In either case, however, the brick must be hard and well-burned. These qualities are necessary for areas where severe winter weather is encountered, along with a lot of thawing and freezing.

The surface texture of brick pavers should not be too rough, especially where foot traffic is heavy or where lawnmowers, wheelchairs, or other equipment will be trundled. Cracks, chips and pits are also to be avoided, as they can trap water and lead to deterioration of the paver surface. In areas where the climate is generally clement, softer brick can be used successfully, but only the hard variety will stand up under constant traffic for long periods of time. By and large, your best bet for paving brick, whether made especially for paving or not, is the SW grade (Severe Weather). In fact, this is the best type of brick to use for all outdoor projects which must stand up to a considerable degree of weathering. Common brick for the most part cannot stand the gaff.

If you should pick out a particular type of face brick rather than standard pavers, take care not to choose those with exceptionally smooth faces, such as glazed brick. These are really too smooth for foot or wheel traffic and can become slick and dangerous when wet, despite the rough joint pattern between them.

Most brick pavers are made in the same rectangular form as standard brick, though there are some special shapes available. Dimensions, however, may vary considerably from those of standard brick. Many pavers are relatively thin, especially those designed for indoor applications or meant to be laid on a fairly heavy concrete base. The dimensions of the paver may also depend upon the type of joint which will be run between the units. Pavers are often laid with their sides and edges touching, with the joint amounting to nothing more than a slight crack. This means that they must be sized so that the width is just half of the length, in order that the pavers can be laid in patterns. Standard face brick, for instance, can only be laid without benefit of a filled joint by arranging them in a stacked pattern where they run side by side and end to end in continuous parallel rows. This is because the overall dimensions are sized to include a 1/2-inch mortar joint. If you attempt to lay them without joints and in patterns, they won't fit together.

Other types of brick pavers are specifically designed to be laid without mortar joints; still other types are sized for a 3/8-inch joint, or even a 1/2-inch joint (the standard-dimension for brick joints). As you can see in Fig. 8-1, there are a number of possible patterns for brick pavers, and your imagination can doubtless suggest numerous others.

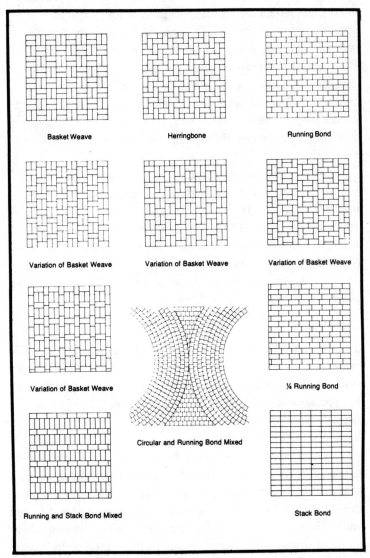

Fig. 8-1. Some of the more popular brick paver bond patterns.

Concrete Pavers

Concrete pavers can be used for the same applications that brick pavers are; the methods and procedures of installation are essentially the same. There are two principal differences between the two kinds of pavers. The material is different, of course, since these pavers are cast from concrete (in somewhat the same manner as concrete block is). The second and most noticeable difference is the tremendous variety of shapes available, with the consequent wide choice of possible patterns. A less fundamental difference lies in the fact that concrete pavers are available in many colors; this makes it possible to vary and enhance the laying pattern—even compose multi-hued mosaics. The coloring agent is mixed with the concrete at the time of manufacture, rather than being applied later, so the color is suffused throughout the paver and will never wear away.

Concrete pavers are manufactured into rectangular shapes—just as bricks are—and into dozens of other plane geometric configurations, many of which are interlocking. There are other types which look a lot like the bottom of an egg carton, and grid pavers full of hollow cores through which grass or other vegetation can grow (Fig. 8-2). Sizes vary greatly, so to determine the number of pavers necessary for a particular job, you'll have to consult the specification sheets of the particular manufacturer.

Laying Pavers

Laying pavers is not a difficult job at all, and as long as attention is paid to a few minor details, an excellent finished product is readily obtained. There are several ways to go about the task, and which one you use is largely a matter of choice tempered by the specifications of the job itself, the amount and kind of equipment you have on hand, and the local weather conditions.

Regardless of weather conditions or the duty requirements of the paved section, best results are obtained by laying pavers on a solid poured concrete base (Fig. 8-3). This base is made in exactly the same fashion as any poured concrete slab with a wood-floated finish. The slab should contain reinforcing mesh or reinforcing rod in the usual arrangement, and in areas of severe winter weather might be further protected by inclusion of a layer of insulation between the

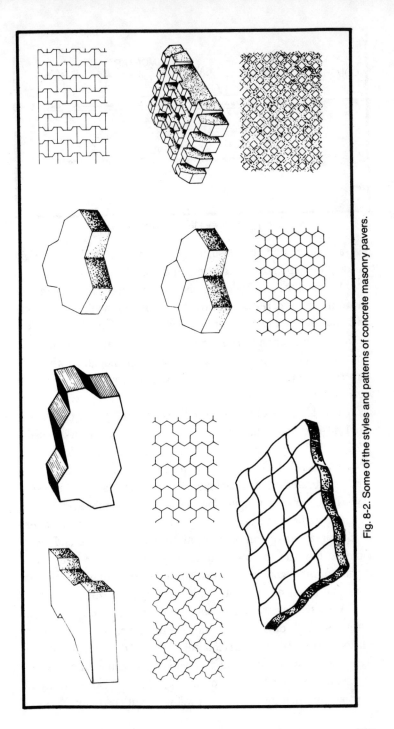

Fig. 8-2. Some of the styles and patterns of concrete masonry pavers.

279

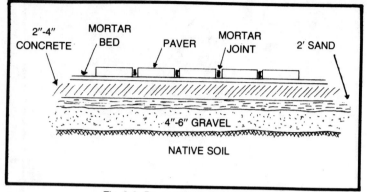

Fig. 8-3. Pavers laid on concrete slab.

slab and the ground. This insulation can be any one of the many rigid building insulations such as foamed plastic.

Forming the slab is done in the usual manner. The side rails of the form may be of redwood or treated wood and should be built level with the finished surface of the paver insulation. This forms a permanent border and an edge restraint for the pavers. Or, the slab could be poured in place on the ground, deep enough so when the pavers are added the final surface will be equal to that of the surrounding lawn or other paving.

The thickness of the slab should be dictated by the conditions. For light-duty applications such as patios, walkways or parking pads, a 2-inch thickness is sufficient if the site consists of well-drained soil and remains relatively dry most of the time. If the area is likely to be wet or saturated often, a minimum 4-inch thickness should be used. If you have a project in mind which might have to handle a fair amount of heavy traffic, especially a long driveway in a cold climate where trucks or a big motorhome might be wheeling back and forth, increase the slab thickness to about 6 inches. An 8-inch slab is capable of withstanding just about any loading or bad weather conditions that might be encountered in residential situations.

After the concrete slab has been allowed plenty of time to cure—usually 7 days or so is ample—the paver laying can begin. Make sure the slab surface is thoroughly free of dirt, debris and dust. If necessary, wash down the slab with a hose and let it dry before beginning. Lay out a few rows of pavers just to check the pattern and to make sure that the dimensions will work out correctly. Now you have a choice to make, because there are four possible methods that you can use to lay the pavers.

The first method employs full mortar joints, but is little different than laying brick in erect structures. Working brick by brick, or perhaps two or three units at a time, throw a bed of mortar on the slab surface and set the first paver in place. Butter the mating side or end, or both, of the next paver, throw a bed and set the paver in place, and so forth. All of the pavers rest upon a full mortar bed, known as a cushion.

The second method is less durable, but will work out all right where the climate is mild and the application is a light-duty one. Eliminate the mortar bed, and set the pavers directly on the dry concrete slab surface. Butter the ends or sides of the pavers, squeeze the units together as you go, and strike the excess mortar from the joints. The joint thickness is usually 1/2 inch, but may be more or less depending upon the desired design and the kind of pavers.

The third method can be used whether the pavers are laid on a mortar cushion or not, but works better with the full mortar bed. Here the pavers are laid down in accurate position, but with a spacing between them equal to the thickness of the desired mortar joint. The mortar used for bedding in this and the two methods already discussed, should be a Type M or thereabouts. After a large section of pavers has been laid, or the job completed if it is not a large one, mix up a batch of loose, soft grout. This should consist of a mix of 1 part portland cement and about 4 or 5 parts fine sand. Add enough water so that the mix is easily pourable. Dump the mix into an old watering pot which has a narrow spout, and pour all of the joint spaces full of grout. This requires a steady, sure, and slow movement of the pouring spout so as not to overfill the cracks or splatter grout about on the paver surface. Keep a sponge and a bucket of clean water handy to mop up the dribbles as you go along. Spilled grout or mortar is difficult to remove after it has dried, and can result in stains which have to be removed with muriatic acid, a miserable chore at best.

The last method is the easiest of all. The pavers are laid out in the same manner as for grout-joint construction, with open joint cracks. Make up a dry mix of 1 part portland cement to about 3 parts of very fine and dry sand. Make sure that the surface of the pavers is completely dry, and spread the dry cement/sand mix on top of the pavers. Carefully sweep the dry materials into the joint cracks until they are filled, then sweep off any excess and discard it. Double-check the paver surfaces to make sure that you have swept absolutely all of the dry mix into the cracks or off and away. If necessary, follow up with a fine-bristled brush such as a dustpan brush. Any cement dust which remains on the paver surfaces can cause stains which will be difficult to remove later.

When all the cracks are satisfactorily filled, spray the whole affair with as fine a mist of water from your garden hose as you can manage. For best results, don't aim the spray directly at the joints, as some of the particles are likely to be dislodged by the force of the water. Instead, direct the mist up into the air and let it fall gently upon the paver surface. Continue spraying until the paving is damp, but avoid any puddling. As the paving begins to take on a dry appearance, spray it again. Continue this process for at least 2 days, and for an additional day or so if the weather is hot and dry. At night, cover the surface with plastic or burlap so that the joints will remain damp.

Once you have decided upon which method of installation to use, assemble your equipment and mix the mortar. Begin at the furthest or most inaccessible point of the project, at one corner or against one side. Lay the pavers in continuous succession from that point to the end. The exact pattern of laying that you follow will depend upon the eventual overall pattern of the pavers that is desired. For some patterns successive full rows will work best, while for others segments or geometric progressions will work best (Fig. 8-4).

If cutting is required, there are two approaches. You can stop and cut the individual pieces as they are needed and as the job progresses, or you can complete the bulk of the job and then go back and fill in the odds and ends afterward. Which method you use depends upon the size of the job and your preferences. Remember, though, that after a short time the mortar will begin to lose its plasticity and will take a set. You will have to get those cut pieces installed before that happens. If you insert the cut pieces as you go along, mortar them in just as you would the full-sized pavers. If you return later to fill in the gaps, treat the cut pieces just as you would

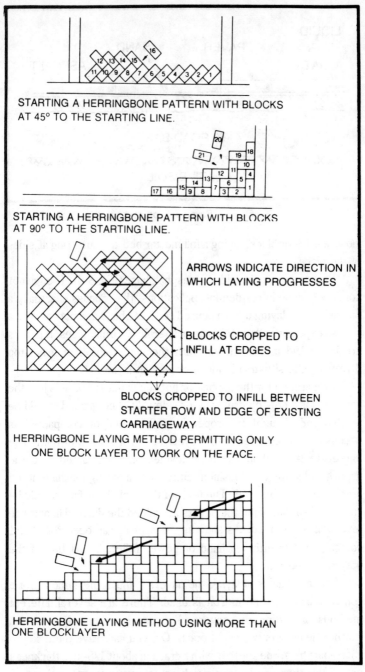

STARTING A HERRINGBONE PATTERN WITH BLOCKS
AT 45° TO THE STARTING LINE.

STARTING A HERRINGBONE PATTERN WITH BLOCKS
AT 90° TO THE STARTING LINE.

ARROWS INDICATE DIRECTION IN
WHICH LAYING PROGRESSES

BLOCKS CROPPED TO
INFILL AT EDGES

BLOCKS CROPPED TO INFILL BETWEEN
STARTER ROW AND EDGE OF EXISTING
CARRIAGEWAY

HERRINGBONE LAYING METHOD PERMITTING ONLY
ONE BLOCK LAYER TO WORK ON THE FACE.

HERRINGBONE LAYING METHOD USING MORE THAN
ONE BLOCKLAYER

Fig. 8-4. Typical paver installation procedures.

283

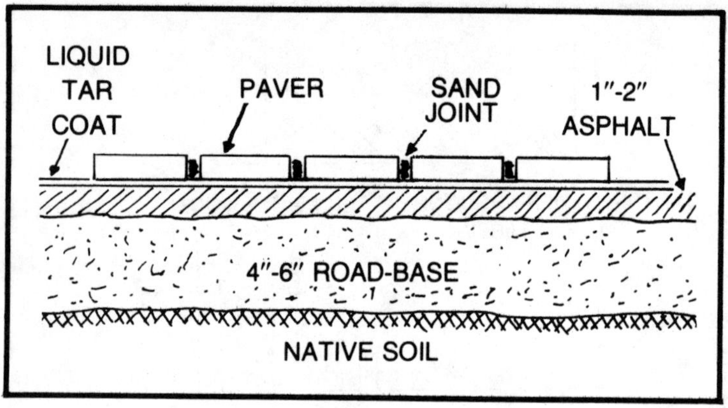

Fig. 8-5. Pavers laid on asphalt.

closure brick or block, laying a full mortar bed and buttering all sides of the piece.

Pavers can also be laid on an asphalt base (Fig. 8-5). Though not commonly done in residential applications, the process is actually a bit easier than laying upon a poured concrete base and may also be a bit less expensive, depending upon specific circumstances. If pavers are being laid in conjunction with the construction of a hot-top driveway, the situation is ideal for asphalt-laying.

Preparation for the asphalt or hot-top paver subbase is just the same as for a hot-top driveway or sidewalk. The ground should be leveled and brought to proper grade, tamped or compacted as necessary, and covered with a layer of sand or gravel, or both. The aggregate layer should also be leveled and compacted, and the asphalt laid in the usual fashion, either with a paving machine, if the area is large, or by hand. The level of the asphalt surface should be calculated so that when the pavers are added the desired final grade level will be attained. As soon as the asphalt has been rolled and compacted with either a hand or power roller, installation of the pavers can begin.

The mechanics of laying the pavers is much the same as for a concrete base, but no mortar is used. There are several different methods of setting the pavers. You can lay them tight to one another, leaving only a slight crack. Or, you can leave them somewhat apart from one another with a crack of about 1/8-inch thickness or so. Whether you leave a crack or not, you can set each paver down dry directly upon the asphalt, or you can swab the bottom of each one with liquid tar (either hot or cold). The latter method glues the pavers to the asphalt base and makes a good solid installation. After all the pavers are in place, dump a quantity of fine, dry sand on the surface and sweep it around until all of the cracks are filled. Follow this up with a light mist of water to help drive the sand down into the cracks, and add a bit more sand if necessary.

On many occasions, especially with light-duty residential applications, the earth itself can serve as a base for the pavers (Fig. 8-6). This immediately eliminates a great deal of cost and labor, and is an ideal situation for the do-it-yourselfer, provided that the ground is suitable. The ground should be solid and well-drained—never marshy or soggy. The less danger there is of frost heaving, the better, and this system works best where the climates are continually mild. On the other hand, frost heaving often levels itself back out, and in any case the pavers can be easily reset or readjusted as necessary. If you don't mind a little occasional maintenance work, earth-bed paver construction will work reasonably well even in areas of severe winter weather.

Drainage of the paved area after construction is also extremely important. The paving should be sloped about 1/8 inch per foot to one direction or another, and the area surrounding the paving should be capable of absorbing and/or shedding moisture runoff.

To make an earth base for pavers, start by digging down below existing grade level to a depth equal to the thickness of the pavers plus the thickness of the sand or gravel cushion which must be installed first. The aggregate (sand or gravel) cushion should be a

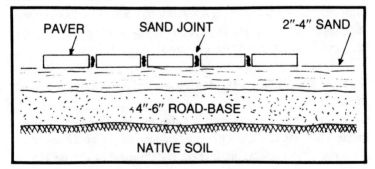

Fig. 8-6. Pavers laid on sand.

minimum of 2 inches thick, and there is no maximum. If the subsoil is hard and has poor drainage capabilities, as much as 1 1/2 or 2 feet of aggregate may be necessary to insure a stable finished area. The worse the natural drainage, the thicker the aggregate layer, but usually 6 inches or so is ample. Do your preliminary excavating carefully so that an undisturbed level surface of native soil is left as a bed for placing the aggregate cushion.

Installations of this type must be bordered with some sort of retaining strip to help hold the edge pavers in place. Otherwise, the repeated action of freezing and thawing, moisture runoff or constant traffic will cause the pavers to creep sideways and be thrown out of line after a time. The edging can consist of 2-inch-thick redwood or other treated wood planks set on edge and staked securely into position. Another common method which produces a good-looking result entails pouring a concrete footing around the perimeter of the area to be paved. As the concrete is poured, a row of bricks standing on end is bedded into the concrete, positioned so that the tops will be level with the finished grade of the paved area (Fig. 8-7). In either case, the dimensions of the space enclosed by the border must be figured very accurately so that the pavers will fit inside with no gapping or lapping over.

With the earth bed and the border finished, the next step is to lay the sand cushion. Incidentally, this cushion can also consist of fine pea gravel, provided that sand joints will not be used between the pavers. This is because the sand in the joints will eventually work its way down into the gravel and disappear completely. With sand joints, always use a sand cushion. Shovel the sand into place and level the surface with a rake. Follow up by soaking the sand thoroughly, and then drawing a leveling board over the surface just as though you were screeding concrete. The soaking will settle the sand grains together and help to compact them. Repeat the process as necessary until you have a tight, level sand surface which extends the same distance below the desired finished grade as the thickness of the pavers.

Laying the pavers is merely a matter of setting them in place one by one, smoothing out the sand cushion as you go if it becomes disturbed from walking around on it. The pavers can be butted tightly to one another, or spaced to provide obvious joints. When the

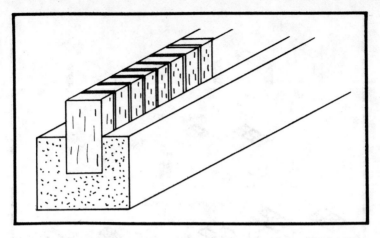

Fig. 8-7. Retaining edge for paved area made by embedding soldier bricks in poured concrete footing.

pavers are all laid, sweep fine sand into all of the cracks and settle it with water.

Pavers can also be used to make level-changing steps as on a patio or terrace. Usually it is possible to work out a combination of riser/tread dimensions (in accordance with the paver dimensions) which will fit the angle of rise of the slope in question. In such installations, however, an earth base is generally not adequate, since erosion, frost action, and repeated traffic can cause the steps to dislodge easily and become dangerous. The paved areas leading up to and away from the steps can be based upon a sand cushion in the usual fashion. The steps themselves, however, should always be mortared to a poured concrete base. The base itself can be poured in forms set upon tamped earth, with the dimensions worked out beforehand to exactly accommodate the pavers and the paving pattern that you plan to use. Form construction, pouring and finishing of this subbase is done in exactly the same manner as for an ordinary set of poured concrete steps, and the finished result looks just the same. The pavers are then set on the concrete subbase in a mortar bed in the same way that pavers are laid on any slab.

CLAY TILE

Clay tile is a masonry building unit which sees a lot less use than many others, for no particular good reason. The fact is, these units

287

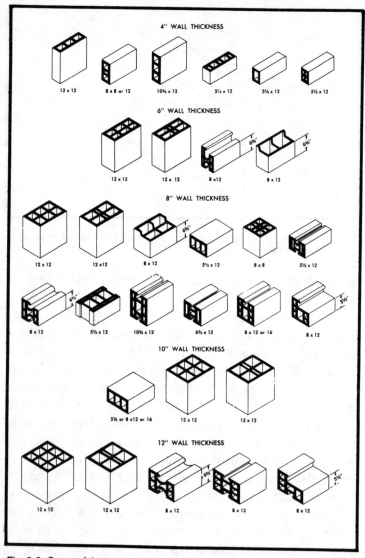

4" WALL THICKNESS

12 x 12   8 x 8 or 12   10½ x 12   5½ x 12   5½ x 12   5½ x 12

6" WALL THICKNESS

12 x 12   12 x 12   8 x 12   8 x 12

8" WALL THICKNESS

12 x 12   12 x 12   8 x 12   5½ x 12   8 x 8   5½ x 12

8 x 12   5½ x 12   10½ x 12   6½ x 12   8 x 12 or 16   8 x 12

10" WALL THICKNESS

5½ or 8 x 12 or 16   12 x 12   12 x 12

12" WALL THICKNESS

12 x 12   12 x 12   8 x 12   8 x 12   8 x 12

Fig. 8-8. Some of the more common configurations of structural clay wall tile.

can produce handsome and interesting effects, are extremely easy to work with, are not especially expensive, and are ideal for do-it-yourself work if you can locate a good source of supply. Clay tile, also often called *hollow tile* or *structural clay tile*, are not flat pieces like ceramic tile, but rather are hollow-core blocks. The material from which they are made is burned clay, similar to brick. Clay tiles are lightweight but strong and contain one or multiple cores. The shells are usually 3/4 inch thick and the webs are 1/2 inch thick. A great many styles and shapes are available, all in modular sizes (Fig. 8-8).

Clay tile can be obtained in two strength classes, one for load-bearing applications and the other for nonload-bearing applications. There are also two basic degrees of quality, with one used for hidden and general-purpose construction work and the other where appearance is of importance. There are a number of surface and face textures and finishes available, including many different glazing styles. Some types of clay tiles have scored or fluted faces on one or all sides, with the indentations serving as keys for direct plaster or stucco applications and for good mortar bonding when the tile is faced with brick or some other type of unit masonry. For home outdoor masonry projects, choose a grade of tile which is made to withstand prolonged exposure to the weather. If the project is a freestanding one which will bear little or no load, use the less expensive nonload-bearing type.

Though there are a good many more or less standard nominal sizes of clay tile, there are also many variations in specific size, design, and particular shape. Before designing a project to be made of clay tile, first check with your supplier and obtain brochures and specifications sheets of the clay tile products that he handles. Design your project around these facts and figures. Check too to see what special shapes might be available that would serve to make the construction job easier or the finished results more attractive or unusual.

Working With Clay Tile

One of the greatest assets of clay tile is its light weight, which makes for easy handling. A 1-cubic-foot section of clay tile weighs about 45 pounds, as opposed to around 125 for brick, and more for stone. Clay tile is a bit fragile, however, despite its great strength as a completed structure, and so it must be handled with care as construction proceeds.

Laying up clay tile follows the same methods and procedures as laying up concrete block, and is nearly the same as laying up brick. There are three possibilities for arranging clay tile in the structure. The first places one tile face to the inside and one to the outside of the structure with the cores running vertically, the same method as is most often used with concrete block. This is called *end construction* or *vertical cell construction*.

The second alternative also presents a face to both inside and outside surfaces of the structure, but the cores are laid horizontally. This is called *side construction* or *horizontal cell construction*. There is no difference in the strength or rigidity of the finished structure with either of these methods, but end construction is a little easier and faster. This is because the mortar is applied to the tile faces, rather than to the relatively narrow shells or webs. End construction, on the other hand, makes it possible to enclose wiring, piping or heat ducting within the sections. In point of fact, both methods are sometimes used concurrently in the same structure.

The third possibility for laying up clay tile is to arrange them with abutting faces, with the cores horizontal and running from front to back of the wall section. This results in a screen wall, which has reduced strength but affords all kinds of decorative possibilities.

As with block or brick, the dimensions and bond patterns must be carefully figured out in advance. The nominal tile sizes allow for a 1/2-inch mortar joint, and the mortaring is done the same way as for concrete block. Once past the first course, which uses a full mortar bed, face shell bedding is all that is necessary. Full bedding can be used, however, if additional strength is required. Reinforcing rods, tie wires, core grouting or filling and anchoring are handled in the same way as for concrete block. Clay tile should always be laid on a solid foundation or footing, with a 1-inch-thick joint between the footing and the first course of tile. Subsequent bed joints and head joints go back to the 1/2-inch thickness. The preferable method of cutting clay tile is with a power masonry saw, since this gives the cleanest cuts and also greatly reduces tile breakage. The alternative method is to score the tile at the breaking point, and then crack or chip off the unwanted piece. This is a tricky business at best, and easily results in a number of tile being thrown away.

CERAMIC TILE

When it comes to outdoor home masonry projects, *ceramic tile* is a building material which is considered only infrequently. Perhaps the most common application is in and around swimming pools, but it is seldom found elsewhere. The fact remains, however, that ceramic tile is one of the best and certainly the most versatile of any of the masonry units, every bit as useful outdoors as indoors, from both a utilitarian and a decorative standpoint. No other type of masonry unit can provide the tremendous artistic possibilities that ceramic tile does. In addition, there are other types which are extremely durable and strongly resistant to weathering, staining and other kinds of damage, with these characteristics often being combined with decorative finishes as well.

Ceramic tile is made in a similar manner to brick and is composed of essentially the same materials. Unlike brick, the units are much thinner and designed to be installed as a veneer and not as a structural element. Primarily ceramic tile is used to cover floors and walls. Particular applications are numberless, and by choosing carefully you can select a kind of tile that can be used for just about any purpose.

There are several advantages to be gained with ceramic tile which cannot be obtained with any other kind of masonry unit. One is the decorative possibilities already mentioned. Another is all-around durability. Colors are not restricted to pastels and earth-tones, but may be utilized in brilliant, eyecatching hues. These colors are fast, and will not fade away from wear or the effects of sunlight. Ceramic tiles are impervious to practically all staining, easily cleaned, and require very little maintenance. Mechanical damage can be corrected by simply lifting out the affected tiles and replacing them with new ones. Some types have great strength and tremendous resistance to abrasion and wear. Design possibilities, of course, are endless and can provide a great challenge to anyone with an artistic bent.

The term ceramic tile is a generic one which covers a great many different types. These types can be broken down first into two general groups; *glazed* tile and *unglazed* tile. Then there are four other classifications which have to do with the *vitrification* of the tile. This is the degree to which the raw tile materials have become fused together by high heat while baking in kilns. An *impervious* tile is one which has either a surface or a body which is so dense that no absorption of any kind of liquid can take place. A *vitreous* tile has a surface or body density which absorbs less than 3 percent of moisture and does not allow penetration of grime. If the tile is *semivitreous*, the absorption rate is more than 3 percent but less than 7 percent. Any tile which will absorb more than 7 percent moisture is called a *nonvitreous* tile.

Glazed tile is available in an awesome number of colors, patterns, and general finish characteristics, as well as in a great variety of shapes and sizes. The glazed finish is on the surface of the tile only, and varies in thickness with the manufacturer. The body of the tile is plain and nondescript. All of these tiles are available in three different general types of glazes which have to do with the light-reflecting properties; *bright* or *glossy, semimatte* or *semigloss*, and *matte*. The matte surface reflects light to a certain degree, but will not reflect an image. A bright surface is highly reflective and will reflect an image. Obviously the semimatte is in between.

Glazed tile are further broken down into several specific categories, many of which are made to exacting standards and are designed for particular purposes. There are far too many of these to investigate in depth here, but an important characteristic to keep in mind is the capability of any given kind of tile to stand up to the weather. This is of no consequence for indoor applications, but is extremely important to the do-it-yourselfer who is designing home outdoor masonry projects. Tile designated as *weatherproof* (with the symbol WP) have been tested by the Tile Industry Research Bureau and found to be able to withstand standard freezing tests without disintegration and thus can be used in outdoor applications.

Unglazed tile has no glazing on the surface and the tile appears the same throughout its entire body. There are numerous sizes and shapes, as well as assorted colors and designs, but the choices are far fewer than with glazed tile. Again, there are many different types, some of which serve particularly well in home outdoor masonry projects. Tile which is either impervious, vitreous or semivitreous can be successfully used in outdoor applications, even where severe weather and freeze-thaw actions are prevalent.

One of the most popular types is *paver* tile, which look somewhat like small brick. Pavers are relatively thin, but are also strong. *Natural clay* tile is somewhat similar, available in mosaic and paver types, and in colors approximately the same as those of brick. *Abrasive* tile is also available in several types, and is impregnated with a small amount of abrasive grain like carborundum. This tile has a high resistance to slipping, and is excellent as a surface for walking. *Quarry* tile is another popular style, presenting a dense and smooth surface in the various natural colors.

Choosing the correct tile for a job requires a bit of research on the part of the designer. Your best bet is to investigate all of the different lines of tile that your supplier has available, and go through all of the pertinent specifications. There are so many variations and possibilities that there is no way to make a blanket statement about what can or can't or should or shouldn't be used, as far as specifics are concerned. Furthermore, the great value of tile to masonry projects derives through its ability to contribute handsome and unique effects and to give opportunity for the expression of imaginative and artistic design concepts. The ideas and the choices should be left to the designer himself.

Setting Tile

Tilework, especially in the more intricate, ornate or exotic designs, is being more and more frequently bypassed in residential construction these days because the cost is so terribly high. That high cost, however, doesn't result from the price of the materials but from the cost of the labor involved in setting the tile. On a per-square-foot basis, uninstalled ceramic tile compares favorably in cost with many other building materials; more than some, but less than others. But the job of setting tile can be a long and tedious one, especially where intricate patterns, mosaics or even scenes are part of the overall design. And of course, the larger the job, the greater the amount of labor involved. But this makes tile-setting an ideal operation for the do-it-yourselfer, because if he is willing to invest the time and labor, he can end up with a classy, rich, artistic and probably unique installation which will greatly enhance his property.

There are two basic methods of setting tile. One is called the *full-mortar-bed* system, wherein a bed of standard mortar similar to that used in other masonry work is laid upon a stable base and the tile is set in the mortar. The second method is called the *thin-bed* system, whereby the tile is stuck to the base with any one of a number of adhesives. Each method has its advantages and disadvantages, but both work equally well and have equal bonding strength.

The full-mortar-bed method is usually preferred for flooring applications, while the thin-bed method is most widely used on wall surfaces. Either one, however, can be used for either application, the choice depending upon the specific circumstances. The thin-bed method is the easiest by far, and ideally suited for the home mechanic.

Ceramic tile in flooring applications should be placed on a solid and substantial base, reinforced to prevent shifting and cracking and heavy enough to withstand whatever traffic loads will be imposed upon it. The base for wall tile applications must also be rugged enough to withstand light lateral pressures, sustain any loading which might be imposed by the structural design, and built solidly enough to preclude the possibility of shifting or settling. In both cases, the base surface must be perfectly smooth and free from any dips, depressions, waves or other irregularities which would prevent the tile from being laid absolutely flat and plumb.

For exterior paving work, the best base is a poured concrete slab, properly reinforced. The tile can be set with either the full-mortar-bed method or the thin-bed method. There are several variations possible for setting the tile in mortar. One is to coat the concrete slab with a layer of mortar from 1/4 inch to 1/2 inch thick, carefully leveling the top of the layer by screeding. The surface texture of the layer need not be smooth, and the finish left by careful screeding is satisfactory provided that there are no ripples or ridges. The surface, however, must be perfectly level. Before the mortar coat is fully cured, but after it has firmed up and is well set, spread a *neat coat* of cement on top. This coating consists of a gruel of cement and water thin enough to pour and having the consistency of a very thick paint. The neat coat serves as the adhesive to secure the tiles to the base.

With the thin-bed method, all that is necessary is to make sure that the slab surface is perfectly clean and dry. The adhesive is then spread with a trowel, usually of the notched variety, in a thin layer over the surface. The tile is set directly upon this adhesive, and the job is done in small sections. There are many different kinds of these adhesives which are designed for various applications, and they are handled in somewhat different ways. Follow the manufacturer's instructions exactly for whatever brand and type you are using. Tile installation procedures may also vary a bit depending upon the particular adhesive. Again, follow the directions given.

Whether you use mortar or an adhesive, the actual setting of the tile requires time and patience and a good eye. Some kinds of tile are laid in sheets, where the tiles are already secured to a backing on which they've been accurately spaced. All the tile setter has to do is align the edges of the sheets. Many other kinds, however, must be laid individually. This means setting each tile in place and accurately aligning it with its neighbors so that the joint spacing is identical, there is no run-out in the joint lines, and none of the individual tiles are crooked or otherwise misaligned. Try to set the tile as accurately as possible on the first attempt, so that there is no need for any more adjusting than is absolutely necessary.

After all the tile is set, or a substantial section if the job is a large one, the next step is to *pound* the tile into place. Go over the entire surface with a block of scrap wood—a chunk of one-by-six about a foot long works well—tapping the wood with a hammer or mallet as you go. There is no need to bash down upon the block as though you were driving nails; a steady firm tapping works best. Pass your hand over the tile surface periodically as you work, looking for bumps or low spots. Iron them out with the hammer and block until you have a perfectly smooth and flat surface. If cement or mortar squeezes up through the cracks between the tiles as you work, clean off the excess with water or an appropriate solvent.

The final step in laying tile is to apply the grout. This can sometimes be done immediately after the tile is set, but in many cases must be delayed long enough to allow the adhesive to cure. The specific time period varies, depending upon the adhesive used, so follow the manufacturer's instructions. The grout may be a fine mortar mix, a premixed packaged grout made up from cementitious material, or an organic or epoxy grout made especially for this kind of work. Which one you should use depends upon your preferences and the conditions to which the tile surface will be subjected. Nowadays the organic and epoxy grouts seem to be the most popular, since they are easily applied, highly durable and not susceptible to staining or moisture absorption. Epoxy grouts are tremendously strong and long-lived and will provide an excellent though slightly more expensive finished job.

The general procedures for applying grouts are roughly the same, though the specific details will vary according to the particular grout and the kind of tile. The pasty grout is first spread upon the tile surface with a trowel or squeegee and worked thoroughly into all of the joint spaces. The joints must be fully packed with material, with no air bubbles or voids and with the top surfaces of the joints approximately even. The top of the joint surfaces is sometimes made almost level with the tile tops, but more often is somewhat indented. Excess grout should be scraped or squeegeed away. The curing time for grouts varies from a few hours to 10 days or so. Following the curing, which may be a dry or a moist process, the tile faces must be cleaned and polished, and any residue removed. The particulars of cleaning are dependent upon the type of grout used.

Setting tile on a wall is done in exactly the same manner. The full-mortar-bed method is sometimes used, starting with a scratch coat applied to a wood or metal lath backing and ending with a plumb coat of mortar and a skim coat of neat cement into which the tile is set. This same system can be used to apply tile to a poured concrete or a masonry unit base. The preferred method, however, is the thin-bed. This involves spreading an adhesive or mastic upon the wall surface, setting the tile into the adhesive, pounding the tile plumb and level, and grouting, just as for a flooring application.

ADOBE

Adobe is the granddaddy of all the premade masonry units, and was the forerunner of brick. The only older masonry unit is stone. As far as modern-day construction in this country is concerned, adobe still makes a marvelous building unit provided that you reside in those areas of the Southwest where the raw material is available and the weather conditions are amenable to adobe construction. Though adobe has enjoyed renewed popularity over the past few years, and can be employed for just about any building purpose, it will only stand up in arid or certain semiarid regions. If adobe is not already in use in your area you have two choices: forget it, or move.

The term adobe refers to material in three different states. The first is the raw material, a native soil or dirt which consists largely of clay with varying quantities of sand mixed in. The second is the same material plus sufficient water to make a heavy paste, which can be used as a mortar or plaster. The third form is the solid adobe dirt block, which makes an amazingly tough and long-lived building unit.

Adobe block or brick is made by scooping up a quantity of native soil, adding water, and sometimes mixing in a quantity of chopped straw for added strength. Other materials are frequently added as well, such as folded tin cans, bottle caps, or anything else that might be handy and serve to tie the mud together. The adobe mud is shoved into a single-brick or sometimes a double-brick wood form and patted smooth. This is usually done by hand and results in a rough and irregular surface texture. When the bricks are partly dry, the mold is removed and excess mud trimmed off and the bricks are stood up on edge in the sun to dry thoroughly. The "standard" size is 10 inches wide by 14 inches long by 4 inches thick and a fully dried brick weighs about 30 pounds. Manufacturing is most often a do-it-yourself project, though there are a few small cottage industries scattered about the Southwest where adobe brick is made to order. Shipping adobe brick more than a few miles from the point of manufacture is not feasible because of handling, weight and cost problems.

Either you like adobe or you do not, and for those who do, there is no better building material. Adobe construction is easy, except for the physical effort required, and extremely forgiving. It is very versatile, and can be built in simple or complex structures. Building experience is not necessary—just a strong back. Adobe can also be used in conjunction with other building materials such as wood and native stone. In many areas foundations are unnecessary, even for a sizable structure such as a home.

Working with adobe block is simplicity itself. Once the project is designed and properly dimensioned, the builder stacks his blocks one on top of the other, usually in a running or common bond of 1/2-overlap, gluing them together with adobe mortar as he goes. Some structures are plastered over on one or both sides with adobe mud as a finish. Others are not. Exact alignment, level, or plumb is often disregarded; this results in a charactistically irregular structure which is so often an integral part of adobe design.

Adobe construction, when properly done, is incredibly rugged and durable, and some structures have been standing for literally hundreds of years. Extremes of temperature have little effect and adobe's only natural enemy is water. In arid areas the material works fine. But in repeated heavy rain, adobe will dissolve.

STONE

Stone has long been an obvious choice for a natural building material and still has great popularity today. The tremendous amount of labor and consequent high labor costs (by comparison with modern building materials) has caused a reduction in stone construction over the past few years, but it is still a viable masonry construction unit, particularly for the do-it-yourselfer. This is especially true where there is an abundance of suitable stone on or near the building site, and/or where the rustic and natural appearance unobtainable with any other kind of building material is desired.

Kinds of Stone

There are many kinds of stone which are suitable for stone masonry, some of which are excellent while others are only moderately successful. There are a few kinds which should be avoided. Knowing which is which is most helpful if you want a solid, long-lasting job.

Rock types can be classified into three large groups, each with its own characteristics; *igneous, sedimentary*, and *metamorphic*. Each of these three groups can then be further broken down into numerous lower categories. Without going into all the geologic details, we'll briefly discuss the main classifications.

Igneous rock was originally formed deep underground by intense heat. The raw materials reached a molten state and then solidified, forming the hardest, toughest and densest of all rock. *Basalt* is one well-known variety, and it is frequently quarried and crushed to be sold as *traprock. Porphyry* and *diabase* are also igneous rocks, the latter also being used for traprock. *Granite* is perhaps the best known of all, and is widely used as a building material either in quarried sections or in natural stones.

The sedimentary rocks started off as layers of silt, sludge and sometimes small pebbles. These materials were carried by river action into large beds which eventually solidified into stone by compression from the weight of the layers above. A good share of the earth is covered by this type of rock, and there are many different kinds. *Conglomerate* rock consists of pebbles, some of them egg-sized or larger, cemented into a solid matrix. This rock is sometimes called *puddingstone* because of its unique texture, and can be successfully used in stonework. *Sandstone* is another common material, and consists of grains of sand cemented together into a mass. This often is a colorful rock; it ranges in hue from white through gray and yellow to dark red. *Shale* is frequently used for stonework, and so is *limestone*.

The metamorphic rocks were originally sedimentary in nature but were also exposed to heat and pressure, partially melted and then reformed. The premier of stone building materials, *marble*, was metamorphosed from limestone. The ever-popular *slate* was transformed from shale. The various *schists* and *gneisses*, as well as *quartzites*, are other examples of metamorphic rock.

Of the three classes, igneous rocks are the hardest to work with. They are difficult to break and even harder to split evenly, and are extremely tough. They don't lend themselves to easy chipping and forming at all. River rocks which have been tumbled and worn smooth are the worst offenders, with glacial till or fieldstone following right behind. On the other hand, this is the most durable stone and in some areas of the country the most common.

Metamorphic rock is relatively softer and more easily worked. It has good breaking characteristics and can be readily chipped and formed, and some types can also be sawed without resorting to the huge milling and sawing machinery found at a granite-works. Most metamorphics are quite weatherable but easy to build with; they can often be obtained in flat plates.

Sedimentary rocks are soft—often too soft to serve as building stone. They are more susceptible to weathering action such as freezing and thawing and wind erosion than the other types. Even so, they have been widely used for all kinds of stonework and can be successfully employed provided that the particular rock is tight and sturdy, and has no tendency to crumble or split apart easily. Sedimentary rock is the easiest to split into slabs, and can be sawed with a power masonry saw.

Stone Construction

Building stone can be put to work in any one of four different states: *dressed, rough-dressed, trimmed,* and *undressed.* On many occasions combinations are used. Dressed stone is cut to regular shape by power equipment into smooth-faced square-edged blocks or plates or whatever other form is needed. Such stones sometimes receive further finishing treatment, like glazing, or polishing with air-driven glass beads or steel shot. Dressed stone is most often used for stone-veneered building walls, paving, or special forms, such as pilasters and columns. The application of dressed stone in home projects most often involves slate or perhaps marble paving units.

Rough-dressed stone is just what the name implies. The material may either be quarried or gathered as free rocks or boulders, or fashioned into rough rectangles for easier laying. The surfaces are extremely rough and often chipped and there is no regularity of size in any plane. Some faces may be angular, and shimming and chinking between the major stones of the structure is the rule rather than the exception.

Trimmed stone consists of chunks, including spoil, just as they are taken from the quarry or as natural rocks and boulders, trimmed and chipped slightly here and there to make them better fit a given position in the structure. This process merely involves chipping off protruding edges or knobs or perhaps breaking off unwanted sections for a better fit. The original shape of the stone is changed little and for the most part not much labor is involved.

Undressed stone is untouched, and used in the same natural shape as it was collected in the field or taken from the quarry.

Not all stones are equally easy to work with in laying up a stone structure. Those which are easiest to lay are roughly rectangular in shape with relatively flat tops and bottoms and rather smooth and flat sides. Elongated plates handle more easily than stubby blocks and are readily laid in a widely overlapping bond for great strength. The most difficult stones to work with are river rocks which have been washed smooth and tumbled nearly round or egg-shaped, so that they stack about as well as a pile of bowling balls. Odd shapes which are essentially curved in all directions and offer few or no flat spots, such as rounded heart shapes, kidney shapes and the like, are also difficult to lay up because they tend to rock and teeter, having no bearing surfaces, and are most difficult to shim or wedge. Obviously, the smoother the rock, the more difficulty you will have in stabilizing it and the more likely it will be to squirt out of the structure. Most stones are neither particularly good nor particularly bad for stonework. The key to good stonework is learning just how to fit all of the stones together into a coherent and cohesive whole—a stable and enduring masonry structure. This is especially tricky with dry-laid stone structures such as walls, so much so that dry stone wall-building has passed into the realm of folk art, and almost a lost art at that.

There are two basic kinds of stone construction; *veneer* and *solid stone*. Veneer construction consists of a relatively thin layer of dressed or undressed stone secured to a base of some sort. The base can be of wood frame construction, as in the case of a building wall, or of poured concrete or masonry unit construction. This method is often used for foundation walls, building walls, columns, and thick-sectioned decorative walls. The method can be adapted to practically any construction project, including a barbecue unit, patio or garden walls, concrete-based stone steps, veneered pumphouse or garden shed, or anything else that strikes your fancy. The whole purpose of employing veneer construction is to reduce the amount of stone needed, as well as costs and labor. However, while the stone used can be of any breadth and length, it must be relatively flat-faced and of uniform thickness. All veneer construction is bonded together with mortar joints and secured to the base with ties.

Solid stone construction is just that, and can be accomplished in any one of three different ways. The first method, *dry-laying*, is especially applicable to freestanding wall structures usually less than 4 feet high, or to retaining walls and laid-in riprapping, which can be considerably higher.

Slip-form stonework is also used primarily for walls, and is a combination of stone and poured concrete. With this method, movable wood forms are made and set in place. The forms must be shallow and heavily braced. A row of stone is arranged in the form and a plastic mix of concrete is poured around them. The exterior face of the wall shows mostly stone, while the interior face is often plain concrete, showing little if any stone. The layers of stone and concrete are built up step by step until they reach the top of the form. The concrete is allowed to cure, and then the form is slipped up to the next level, and the process begins again. This makes an ex-

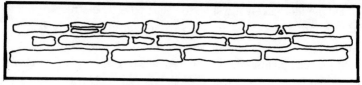

Fig. 8-9. Ashlar stone construction.

tremely solid wall, and contains approximately equal parts of stone and concrete.

Laid-up stone can be used to form any kind of structure, including walls, steps, barbecues, columns, and gateposts. This method involves laying each stone in place in a mortar bed, carefully fitting and arranging the pieces during the process much in the manner of laying brick or block. As with all stonework, careful fitting together of the components is necessary to make a good job. Extra care has to be taken if the mortar joints are to be narrow and uniform. If this is not a critical point, the fitting together can be less precise.

Whether the stonework is solid or veneer, the construction can be classified into two different types; *ashlar* or *rubblestone*. If the stones are flat and of a more or less rectangular shape—like elongated, outsized bricks with irregular margins—and are laid lengthwise with considerable overlapping, this is ashlar construction (Fig. 8-9). Where the rocks are chunks and splits of varying sizes and shapes, having only one or two flat surfaces for the most part and irregular in shape, this is rubblestone construction (Fig. 8-10). Such stones may result from breaking larger rock, may be the by-product of quarrying operations, or can be natural stone as found.

An offshoot of rubblestone construction is called *fieldstone* work, where the stones used are glacial till, frequently chunks of granite. The stones are laid in whatever shape they are when collected from the fields, except perhaps for a bit of trimming here and there. Another offshoot of rubblestone construction, with its own particular appearance, is *river rock* or *cobblestone* construction, consisting of relatively small and roughly spherical, smooth stones obtained from river beds and banks. This type of construction makes heavy use of mortar, and is often used as a veneer. It is also popular in many areas for assorted garden and yard projects.

Mossrock construction is essentially a variant of fieldstone work, where the object is to choose stones which have one face

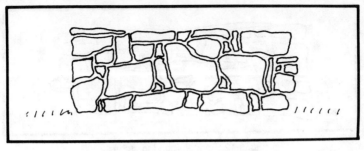

Fig. 8-10. Rubblestone construction.

liberally covered with mosses and/or lichens (also known as *lichen-rock* construction) for decorative purposes. This rock is left in its natural shape and is carefully fitted and interlocked; it must be handled with extreme care during the construction process so as not to disturb or destroy the growth. *Flagstone*, another often-heard term, is a variation of ashlar construction, employing long and flat slabs from 1 to 4 inches thick in an overlapping pattern.

There are two bond patterns common to stonework; *random* bond and *course* bond. The course bond is most often found in ashlar or flagstone structures, where the regularity of the stone thickness lends itself to laying out the pieces in relatively straight rows or courses with fairly straight horizontal joint lines that are obvious to the eye (Fig. 8-11). Rubblestone construction can also be done this way but is more difficult and the horizontal joint lines are not likely to be as straight. The random bond is the more common in stonework, with the joint lines following the contours of the individual stones as they are fitted into the structure (Fig. 8-12). This produces a series of wandering and asymmetrical joint lines which follow no regular pattern at all.

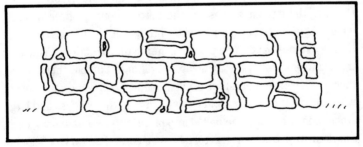

Fig. 8-11. Course bond in stone.

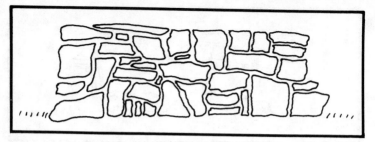

Fig. 8-12. Random bond in stone.

Working With Stone

There are two main points you should understand about working with stone. The first is that it is a heck of a lot of hard work. The second is that you should always try to make things as easy on yourself as you can. Work methodically and pace yourself, and never strain unnecessarily. Stone is not only sharp, rough and unwieldy, it is also very heavy. To give you an idea, sandstone (one of the lighter ones) weighs in at about 150 pounds per cubic foot—heavier than poured concrete. Granite runs about 170 pounds, slate about 175 and basalt about 185. As you can see, it doesn't take a very big rock to approach the 100-pound mark; it's dead weight in a small package and tough to handle.

When lifting stone, never place the strain on your back. Keep your back as straight as you can, hold the rock in close to your body, and lift with your arms and legs. If you get the wrong grip, release immediately and try again. If the rock begins to slip or you get off balance, drop the thing and get out of the way. A wrong movement or a mishandled stone can lead to agonizing muscle pulls or even ruptures and other assorted dreary ailments.

Never try to pick up more than you can comfortably handle. Reserve the largest stones for the bottom course of the project, where they can be rolled or levered into position with bars or poles. If heavy stones have to be placed higher in the structure, maneuver them into position without doing any lifting. Tumble them or flip them end over end up stout planks; use inclined planes and rollers whenever you can. Bars and levers can help the process and large stones can sometimes be pulled into place with a stout logging chain attached to a truck or tractor. In that case, be wary of slipping chains.

The easiest way out, of course, is to stick with small stones which can be moved about without too much effort. Depending upon what kind of a stone supply you have, this sometimes necessitates making little ones out of big ones. The easiest way to accomplish this chore is to "have at" the big ones with a sledgehammer with as heavy a head as you can reasonably manage. Most rocks have at least one crack in them somewhere, or one or more cleavage lines along which they will readily break apart. These points will sometimes be obvious, and sometimes not. Once you locate the right spot a few healthy shots with the sledge should break the rock. There are stones, sometimes referred to as *hardheads*, which simply refuse to break. Don't waste your time on them because they are just too tough to break by hand. Wedges or wedge-and-shim sets are often helpful in breaking up stone, provided that you can find a crack or fissure in which to get the wedge started. Beware of flying wedges and shims.

It happens with dismaying frequency that the stone you are trying to work into a particular slot almost fits, but not quite. It may have a projecting corner, a wobble knob on the bottom, a straight edge that needs a slight curve, or some other such problem. Or there may just be protruding edges that you don't like the looks of; or perhaps you would like to rough dress the face. All of this means that a little trimming must be done.

Depending upon the specific circumstances, you can do this with a Bush hammer, a stonemason's hammer or a brick hammer, a geologist's pick, or a hand sledge and stone chisel. Trimming with picks or hammers should be done with a series of stout raps, rather than bashing away at the stone. Flake off the unwanted material bit by bit, so that you don't run the risk of breaking up the stone. For removing larger chunks, the sledge and chisel combination works best, especially if you can locate some cleavage lines or handy cracks. Score the entire cutting line first with the chisel, with fairly light taps. Then go back along the line with stouter blows. If the pieces don't come free, rap the waste portion with the sledge itself. Repeated attacks with chisel and then sledge will usually turn the trick. If the chisel blade or hammer strikes sparks from the stone, you may have a problem. You might have hit an isolated hard spot in the stone, in which case it will probably split eventually. But you might also have an extra-stout chunk of igneous rock which won't split no matter what. If that's the case, use the stone somewhere else, whole.

Laying Stone Dry

Laying up stone without mortar is the traditional method, and nowadays is used principally only to make stone walls. And there is little doubt but that a stone wall properly and expertly laid without mortar will stand firm and square for decades, long after a mortared wall cracks and begins to crumble. There are a lot of tricks to building a fine wall, and the know-how and the "eye" for stone placement comes only through experience. There are stonemasons who have raised the craft of building stone walls to an art form, and watching them work gives one pause to wonder how in the world they put everything together so handily. But even the totally unexperienced do-it-yourselfer can turn out a creditable job if he keeps a few points in mind.

Dry-laid stone walls can be placed directly upon the ground, provided that the total height of the wall does not exceed 2 or 3 feet. However, the best method, regardless of height, is to sink the bottom course or two into the ground. Dig out the vegetation and topsoil until you strike a layer of compact and dense subsoil; this will form the base for your wall. Set up stakes and guidelines so that you can keep everything square and true.

The wall should be at least two rows of stone in thickness. Lay out the first course by choosing the worst, largest or least workable of your stones and bedding them firmly into the subsoil. Nest and fit them together, with the best sides facing upward, and pack soil in and around them to hold them in place. The upper faces of the stones should never cant to the outsides of the wall, but either be perfectly flat or slant inward toward the centerline of the wall. This slant need only be at a slight angle, and is carried through for the full height of the wall so that the two rows lean against one another and are self-supporting, with no tendency for any of the stones to slip to the outsides of the wall.

From here on, laying up is a continuous process of fitting and adjusting the stones in either random or ashlar bond pattern. The trick is to nest and fit them together tightly, so that they do not rock or wobble and are solidly bedded. The general principle to follow is "two on one, one on two." This means that one stone should lie more or less centered upon the joint between two stones below, with two more overlapping above. This obviously is not always possible, but makes for the strongest wall and is the goal to shoot for.

Use as little chinking with small flat or wedge shaped stones as possible. The more regular in shape the principal stones are, the less need there will be for chinking. Obviously, though, irregular shapes are more the rule than the exception, and so it is extremely difficult to get by with no chinking at all. Chinking and shims should be arranged in such a way that the principal stones act to wedge against one another and toward the centerline of the wall, rather than toward the outsides of the wall. The shims and chinks themselves must also be set so that they cannot slip free.

The weight of the stones and the action of gravity, plus the interlocking of the units, is what holds a dry-laid stone wall together. Gravity acts to pull straight downward, which means several things to the wall builder. First, the wall itself should be built on a longitudinally level line. Rather than following sharp contours of the ground, undulating up and down, keep the courses and the top of the wall level. This eliminates any tendency for the stones to sideslip. Where changes in elevation are necessary to follow the slope of a grade, take the wall up in steps rather than sloping curves. Both faces of the wall must be either perfectly vertical or slant back slightly toward the centerline of the wall. If the sides of the wall slant outward, the overhanging weight has a tendency to pull the wall apart. Ends and corners should be perfectly vertical, or as much so as is possible. Always place each individual stone in the structure in such a way that it will lock with its neighbors and also have the full force of gravity working upon it to help tie the wall together.

As the job goes along, search out from your collection of stones some of the longer and more rectangular specimens. Insert these stones every few feet in every course so that they run all the way through the wall from one side to the other. Tie-stones help to further lock the structure together, and should be in staggered locations from course to course so that one is never directly above another. For most walls, one tie-stone about every 4 to 6 feet is ample.

The best rocks should be saved for the top course, the cap. If possible, they should be large enough to reach across the wall. If not, they should be arranged in staggered pairs, sloping slightly inward to the centerline of the wall. Needless to say, these cap stones must fit just so and with joints as tiny as you can manage, so that the wall top is solid and secure. Remember, there is nothing there to hold them except gravity and their own weight.

The side of the stone which faces the outside of the wall is the most important one as far as appearance is concerned. Keep this in mind as you position the stones and fit them together so that the squarest or most attractive face lies outward. They should for the most part be relatively large ones, perhaps from time to time interspersed with 2 or 3 smaller ones in a cluster. All of these facing stones, however, should lock tightly together so that there is no chance for them to slip out. Avoid the use of small stones in the wall face, since they probably won't stay put. Instead, set them into position on the inside of the course row to fill the gaps and help wedge the face stones in place. Where you are short of stone, but the ones you do have are flat and slab-like and (preferably) fairly good-sized, you can build the faces from these stones and fill the center of the wall with small stones and gravel. Tie-stones should be placed in the usual fashion to help lock the two faces together in good fashion. A wall built in this manner will be somewhat less sturdy—it may only last 200 or 300 years—but is a perfectly legitimate method of construction.

Laying Mortared Stone

Dry-laid stone structures are in a constant state of movement, shifting a bit, settling a little. Because of the way in which they are laid up (if done properly) and because there is no mortar, the structure is not rigid and cannot crack apart or crumble. The situation is different, however, with mortared stone structures. With all of the units tied solidly together the structure becomes inelastic and inflexible, and if not properly based will eventually crack apart. This means that all mortared stone structures must be built on a substantial footing of poured concrete. The dimensions of the footing are calculated in the same manner as those for other types of unit masonry construction, and should be placed 2 feet deep in the earth or below frost line, whichever is deeper. Reinforcing is essential because of the heavy weight piled on top of the footing and also to prevent shifting and cracking.

Once the proper footing is built, laying up a mortared stone wall follows in much the same pattern as laying up a dry stone wall and also uses many of the techniques of other types of mortared masonry unit construction. The type of mortar to use is determined in the same way as for block or brick, but the quantity needed for a given volume of stonework is likely to be at least twice as much as for other types of masonry construction. This is because the units are so irregular, and all of the crevices and voids must be thoroughly filled.

As with any construction of this type, the first course is extremely important. The stones must be laid in a deep full mortar bed, thick enough so that all of the voids between the footing top and the irregular stone bottoms will be completely filled. All of the stone should be free from dirt or dust, and should be dampened before laying. This is particularly important with porous stone like sandstone, which should be well soaked.

The actual fitting and laying should be done in the same manner as for a dry-laid stone structure. Let gravity and weight work for you, with the mortar serving as a tie. The mortar will be the weakest part of the structure, and should not be relied upon to hold poorly fitted stones in place. The better the stones interlock and fit together when dry, the stronger and less susceptible to cracking the structure will be.

After a small section of the first course is laid, smooth the mortar around the stones and fill in any gaps or voids. Then start the second course, progressing in slanted sections. Fit each stone into place so that it seats securely, remove it and remember its exact position, and spread a suitable dollop of mortar into the slot. Set the stone back in its original position and push it firmly home. Again, tuck mortar around the stone and into the crevices as before. As the job progresses, pause to strike the face joints in whatever manner seems appropriate.

After the mortar joints become thumbprint hard, tool them with a pointing trowel or a jointing tool to make a dense and weathertight surface. In areas of high rainfall and/or bad winters, the best bet is to make those joints deeply inset and slanting down and out from top to bottom. The overhanging stone will afford some protection from moisture, and the slant of the joint will quickly shed mosisture which might otherwise accumulate. Cavities and voids between rocks, especially those to the inside of the wall, can be filled with whatever amount of mortar is necessary. Small stones and pebbles can be mortared into place in these spaces to save mortar and further strengthen the wall. However, the joints between the bearing faces of the principal stones, and also the joint lines at the faces of the structure, should be kept thin. A 1/2-inch joint is fine and there will be places where the joints will inevitably be thicker or thinner. But the less mortar exposed to the weather, the more durable the structure will be. Incidentally, it is a wise idea to remove excess mortar from the stone faces with a wet sponge as the job proceeds. Dried mortar stains are especially difficult to remove from stone.

Stone Veneer

Veneering with stone is done in almost exactly the same fashion as veneering with brick. The stone units are tied to the substructure with metal tie-straps or similar means, embedded in the mortar joints and attached to the substructure. Mortar requirements and construction procedures are the same. Probably the greatest difference lies in the fact that the stone units must be trimmed, cut or joined with great care so that they fit together with relatively uniform joints and with a minimum of small pieces visible. The stone itself must be carefully selected so that all pieces are of approximately the same thickness and have fairly flat front and back faces. The thickness chosen generally ranges between 3 and 6 inches. Either random or ashlar bond patterns are used. Native stone in random sizes and sometimes even cobblestone is often employed in veneer construction. However, there are also many varieties of precut and milled stone slabs available expressly for veneering purposes. Such stone is often favored because it is easier to lay up and generally costs less in the long run. However, this type of veneering seldom has the natural rustic charm that many people prefer. Extra care must be taken with mortaring and joint tooling in mossrock and lichenrock construction to avoid unsightly mortar blemishes which are nearly impossible to erase.

9

Walls, Retainers and Fences

Among the most popular of all outdoor home masonry projects are walls of various sorts. Many of these are of the freestanding variety, while others serve as retainers or full retaining walls. Such walls may actually hold back and bulkhead bankings or earth-fill areas, or may act as abrupt level-change controls, as in terracing a slope. Masonry fences may also be combination fence-walls, and serve mostly as boundary markers and adjuncts to the general landscaping design of the home grounds.

There are literally thousands of ways to go about building walls and fences, and many materials and combinations of materials from which they may be built. The possibilities are bounded only by your imagination and skills. The few projects that follow are examples of some easy ones which you can undertake yourself with no difficulties. Hopefully they will also act as springboards to launch you into a few projects of your own design.

FREESTANDING DRY STONE WALL

Lay out the course of the wall and drive some grade stakes to help keep you lined up as you work. Then dig a trench somewhat wider than the wall will be. About 18 to 24 inches is a good wall width, though you can go wider if you wish. Anything much narrower than 18 inches is not likely to work out too well unless the stone is fairly

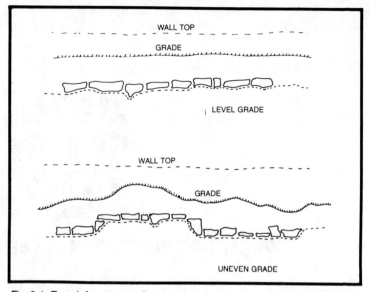

Fig. 9-1. Trench for stone wall on level grade is essentially flat-bottomed, with stones pocketed into soil as necessary (above). On rough grade, trench is stepped with each section essentially flat-bottomed.

sizable and in large slabs or plates. Just below the topsoil is usually deep enough, though you can keep digging down to below frost level if you want to. Keep the bottom of the trench flat. If the ground is level, all you will need is a simple trench. However, if the ground is irregular, terrace the trench bottom into a series of flat planes which roughly follow the ground contour (Fig. 9-1).

Begin the first course by laying out a double row of bottom stone. Pick out stones which have marginal usefulness, such as large ones which must be levered into place, rounds, hardheads, and irregular shapes with poor faces. Nest these stones into the ground, digging slight holes and pockets for them as necessary and fitting them closely together. Pack earth around them so that they are immovable. Try to set them so that their tops are on a roughly level line, and so that the top faces slant slightly inward to the center of the wall. If you can reach the site with a garden hose, sprinkle a little water on the earth so it'll tamp more solidly into place around the bottom stones.

If more courses are necessary to build up to the grade level, continue to use the poorer stones, fitted and shimmed as necessary with earth rammed into all the cracks and crevices. Place the best stones at the end or corner of the wall, locking them together as tightly as possible. The wall end is subject to all sorts of peculiar stresses and strains, and should be put together in a substantial manner. As many stones as possible should run the full width of the wall and be flat enough or interlock well enough so that they are firm and secure without benefit of shims. Once the wall structure reaches grade level, pause to backfill whatever remains of the trench opening. It is a good idea to ram the earth down solidly with a post or a tamper. Spraying with water will give you better compaction, provided you don't turn the soil to mud.

The above-ground courses can now be laid up. Fit all the stones together as tightly as possible. Some shims will doubtless be necessary, but keep them to a minimum. Turn the stones over and around and about, trying for the optimum fit. If one stone does not fit well, discard it and try another. Every 3 or 4 feet, set a tie-stone which fits crosswise the full width of the wall and locks the inner and outer courses together. As with the bottom course, subsequent courses should slant in slightly toward the centerline of the wall. From the standpoint of appearance, as well as strength, as much shimming as possible should be done from the inside of the wall rather than the outside faces. The same holds true for chinking, though there is seldom much chance of getting away from either entirely.

As you lay up the courses and sort through your supply of stone, set aside the best ones to use later as cap stones. Slabs and plates of a roughly rectangular shape usually work best for this purpose, and the larger and heavier (up to a point) the better. Pause frequently as you work to sight down the wall faces to make sure that the structure is staying in a straight line, and that none of the stones are protruding overmuch from the wall. If the wall is a low one, the sides should be approximately vertical, but never slanting outward. The taller the wall, the greater amount of *batter* each face should have. Batter is the degree to which each face leans inward from the vertical. A 6-foot wall, for instance, might be as wide as 4 feet at the bottom and taper to only 18 inches or so at the top. Keep a sharp eye on the level of the wall top. As you approach the final height of the wall, begin to fit the cap stones into place so that the top of the cap course will be flat and level the full length of the wall. In cross section, the wall will look something like the sketch in Fig. 9-2.

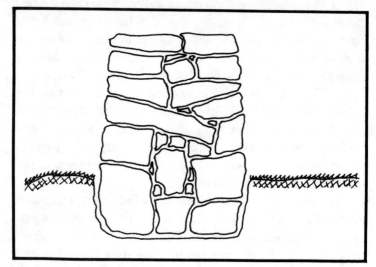

Fig. 9-2. Cross section of freestanding stone wall.

DRY STONE RETAINING WALL

A dry stone retaining wall is put together in much the same manner as a freestanding wall. There are, however, a few differences in overall design. Chief among them is the fact that the face of the wall should have plenty of backward slope, leaning into the earth behind it. The taller the wall, the greater the batter, or backward slant. The inner face of the wall can be built perfectly vertical, and this is generally the case when the wall is built first and fill dirt is placed behind it afterwards. However, the construction is simpler if the wall can be built against a slightly slanted cut-blank face. As the wall is constructed, the courses are tilted back into the bank with the rear row surrounded by rammed earth packed between it and the existing bank.

Drainage is another important factor in the construction of dry stone retaining walls. In areas where ground moisture or precipitation is plentiful, the rear row of stone should be packed with gravel, both around the stones and between the back row and the banking itself. Natural weep holes can be provided by leaving occasional large cracks open between stones. These weep holes should be positioned every few feet along the base of the wall. If the wall is a tall one, provide more weep holes 2 or 3 feet below the wall top. Drainage pipes can be built into the wall to serve the same purpose.

To build a dry stone retaining wall, start by digging a trench about 2 feet deep along the edge of the banking to be retained. The width of the trench is dependent upon the height of the wall. About 18 inches is sufficient for a low wall, expanding to 4 feet for a 6- or 7-foot wall. Lay up the stone in the same manner as for a freestanding wall, packing the bottom courses and the rear rows of stone with earth or gravel. Set the weep holes or drainage pipes as you go along. Slant the face of the wall back into the banking by setting each course back a little bit. The wall cap course may be level with the existing grade at the top of the bank, or the wall may be carried above grade. The cross section of a finished dry stone retaining wall appears much like the one in the sketch shown in Fig. 9-3.

MORTARED STONE WALL

A mortared stone wall should be built upon a firm foundation of poured concrete. The footing can be made level with the existing grade or positioned above or below grade. The bottom of the footing should extend below frost level. If the footing is below grade, concrete block can be used to reach grade level if stone is in short supply.

Rubblestone or fieldstone walls, whether of random bond or course bond, are laid up in the same fashion as a dry stone wall

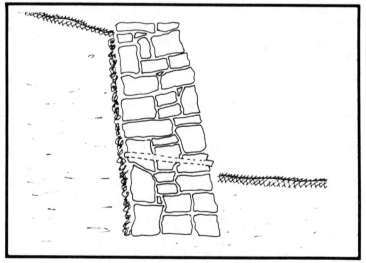

Fig. 9-3. Cross section of stone retaining wall.

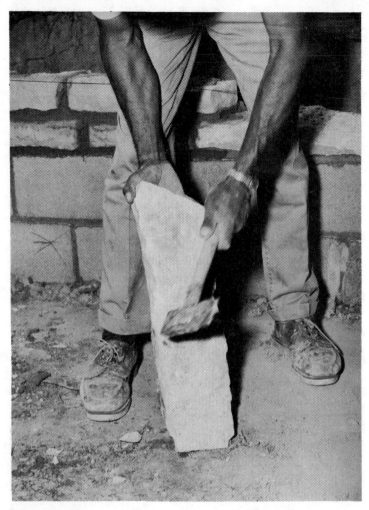

Fig. 9-4. Knocking protrusions and wobble-knobs off flagstone prior to laying in wall.

except that the voids and cavities between the stones are filled with mortar. Careful fitting of the stones is important, though the mortar takes the place of shimming and chinking. Small stone, even only fist-sized ones, can be successfully used in a mortared stone wall. Also, the thickness of the wall can be as little as 8 inches or so for low structures.

Building with flagstone or with other stones of a somewhat regular shape is much easier. Lay a full bed of mortar about 2 inches thick along the footing top and set the first course of stone just as you would brick. Set up a mason's taut line for the second course to give you a guide for leveling. To start the next course, pick a flagstone and knock off any bumps or protrusions which might interfere with the laying (Fig. 9-4). Throw a full bed of mortar in place and set the stone. In this case the stones are in a running bond, and the overlapping can be adjusted to suit. The more overlap there is, the stronger the wall. As you set each stone, tap it into place with the trowel handle until it is even with the taut line (Fig. 9-5). Check the level of each course frequently with a spirit level to make sure everything is properly lined up (Fig. 9-6). Scrape excess mortar away from the joints as you proceed. Since the stone faces are so irregular, the struck joint faces will be likewise. As soon as the mortar sets thumbprint hard, go back over the joints with a pointing trowel or a jointing tool and compress the mortar into a dense joint. In walls exposed to the weather, the joint shape should be concave and carried well back beyond the edges of the stone. After tooling, use a stiff-bristled brush to thoroughly clean out the joints (Fig. 9-7).

Fig. 9-5. After laying a full bed of mortar, flagstone is set roughly level with taut line.

Fig. 9-6. Flagstones are leveled out in mortar bed by tapping with trowel handle.

If the wall is a freestanding one, it needs no further attention. If a retaining wall, however, care should be taken to allow the joints to air-cure for several days before any lateral pressure is placed against the wall. For best results, moist-cure the joints by keeping them damp with a water mist for at least 3 days. After the mortar has cured for at least a week and preferably longer, fill dirt can be placed between the wall and the earth bank which it is to retain. As with the stone wall, if ground moisture or heavy precipitation is a factor, weep holes or drainage pipes should be installed in the wall as construction proceeds. The space between the wall and the banking should be filled with fairly fine gravel up to about a foot below grade level. Topsoil fills the remaining space. If drainage is no problem, throw spoil dirt that is free of rocks and debris into the cavity between the wall and the banking in layers. Tamp each layer firmly and soak with water just enough to settle the soil.

SOLID CONCRETE BLOCK WALL

A solid concrete block wall is one of the easiest of all to build. Standard 8-inch by 8-inch by 16-inch concrete blocks make a structure of ample strength for any residential purposes, although a thicker wall can as easily be built, and thinner ones are also satisfactory for some purposes. Figure 9-8 shows a typical solid concrete block wall construction.

Start the proceedings by pouring a suitable concrete footing, at least 18 inches deep but in any case below frost line. Reinforce the footing in the usual fashion for concrete block walls. If the wall is higher than 4 feet, extend reinforcing rods up out of the footing to a height of 4 feet or more, depending upon the finished height of the wall.

Allow at least 3 or 4 days for the footing to cure. Then lay up the concrete block in consecutive courses until the finished height is reached. If the wall is reinforced, pour grout or pack mortar into each core series through which the reinforcing rods run. Finish the wall by laying a course of cap block along the top. Allow below-grade mortar joints to cure before backfilling the footing trench with earth.

The solid concrete block wall can be made from standard block or pattern block, and can also be treated with a number of finishes.

Fig. 9-7. Joints are tooled and then brushed free of mortar crumbs.

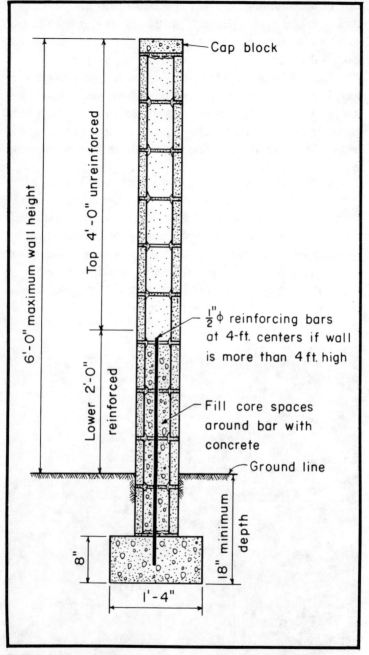

Fig. 9-8. Cross section of a typical concrete block garden wall. Courtesy of the Portland Cement Assoc.

For instance, the wall could be covered with ceramic tile or veneered with stone. Painting is another possibility, or the blocks themselves could be colored, either with one color or with two or more contrasting colors in a pattern. Solid block can also be staggered or offset slightly in their courses, then covered with a layer of stucco to present an appearance something like the one shown in Fig. 9-9.

CONCRETE BLOCK RETAINING WALLS

A concrete block retaining wall is laid up in the same fashion as a freestanding or structural concrete block wall, but there are a few added features. If the wall is a tall one, thicker concrete blocks are frequently used in order to withstand the lateral pressures thrust upon them. The footing may also be wider or thicker than normal. Drainage channels or pipes can be set along the base of the wall and frequently a back drain is set along the base of the wall in a gravel pocket running the full length of the wall. The back-drainpipe may simply run off to either side of the wall, or can be connected to drainpipes which go through the base of the wall. Figure 9-10 shows a typical concrete block retaining wall.

Fig. 9-9. Stuccoed masonry garden wall.

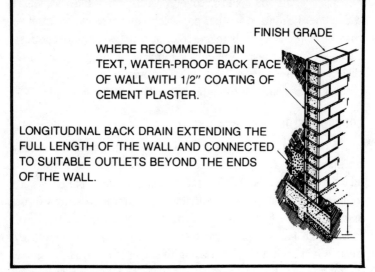

WHERE RECOMMENDED IN
TEXT, WATER-PROOF BACK FACE
OF WALL WITH 1/2″ COATING OF
CEMENT PLASTER.

FINISH GRADE

LONGITUDINAL BACK DRAIN EXTENDING THE
FULL LENGTH OF THE WALL AND CONNECTED
TO SUITABLE OUTLETS BEYOND THE ENDS
OF THE WALL.

Fig. 9-10. Details of a typical concrete block retaining wall. Courtesy of the National Concrete Masonry Assoc.

Such retaining walls are frequently waterproofed on the back face to preserve the integrity of the joints and to prevent moisture from seeping into the cores of the block wall. One good method is to paint one or two coats of waterproofing tar onto the structure. A better though more difficult method is called *parging*. This consists of coating the entire back face of the wall and the footing top as well with a layer of mortar or a plaster made from portland cement (Fig. 9-11). The coating can be applied with a trowel in one 1/2-inch-thick layer, or in two 1/4-inch-thick layers. If two coatings are used, the first should be roughened up with a scratcher after it has partially hardened, kept damp for 24 hours and then coated with the second layer. The wall should be dampened slightly before application, and the plaster coat kept moist for at least 2 days to allow proper curing.

CONCRETE BLOCK SCREEN WALL

A concrete block screen wall must always be built upon a solid footing. The dimensions of the footing should be sized according to the block dimensions in the usual fashion, provided that the wall is comparatively long and high. For small sections of screened walls, the method shown in Fig. 9-12 is satisfactory.

Fig. 9-11. Parging a concrete block wall.

Once the footing is properly established, all that remains is to stack the screen block on edge with a 1/2-inch mortar joint between them. Some types of blocks may have to be laid in a certain direction in order to gain the full pattern, so be careful to get them properly arranged. The screen wall shown in Fig. 9-13 is set upon a course of standard block which will be hidden when the grade is brought up to finish level. The blocks are set between end posts anchored in the footing, making alignment of the wall easier and providing a certain amount of support and stability. Lay the blocks up carefully and keep them perfectly aligned, being guided by a mason's taut line (Fig. 9-14) as well as a spirit level and rule. Note the use of ladder-type

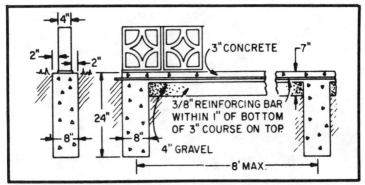

Fig. 9-12. Diagram of footing for a small concrete grille block wall.

Fig. 9-13. Setting second course of concrete grille block wall. Note concrete block starter course to bring grille blocks up to grade level. Courtesy of Sakrete, Inc.

Fig. 9-14. Setting mason's taut line to guide next course of grille block. Note reinforcement lying on block tops. Courtesy of Sakrete, Inc.

reinforcing to provide extra strength. The completed wall, really a rather simple project with great effectiveness, is shown in Fig. 9-15. There are plenty of possibilities for this type of wall, and they can be made curved as easily as straight (Fig. 9-16).

Another idea along these lines provides a similar effect but is made in an entirely different way. This is the louvered wall shown in Fig. 9-17. The first step in building this type of wall is to construct a substantial footing of poured concrete, the top of which lies 3 feet below grade level. On top of the footing, lay a single course of 8-inch by 8-inch by 16-inch trough or lintel blocks. Cast a series of reinforced concrete louvers of appropriate length. Stand the louvers on end, well-supported and properly aligned in the hollow concrete blocks. Pour concrete or mortar all around the louvers and let the whole affair cure for several days before removing the supports. Backfill over the footing and base blocks and tamp earth in firmly around the louvers. The louver tops can be left open, or may be capped with block mortared into place. In the latter case, anchoring

Fig. 9-15. Completed grille block screen wall.

Fig. 9-16. Curved grille block screen wall in a different pattern.

lugs should be cast into the louver ends to help secure the cap course. The louvers can be made of colored cement, coated with cement paint, or left in natural gray or pure white concrete color.

BRICK RETAINING WALL

Brick retaining walls must be built substantially if they are to be effective and long-lived. The wall diagrammed in Fig. 9-18 is a particularly strong 2-layer or 8-inch wall which, while certainly not a weekend project, is not terribly difficult to build.

Dig a suitable trench at the base of the bank to be retained and pour a footing as shown in the diagram. Wire all of the rebar sections together, and don't forget the upright bars which help withstand lateral pressure and add a considerable amount of strength to the wall. Let the footing cure for a least a week and preferably longer before beginning to lay up the brick.

Lay up each course of brick in sequence, throwing down a full mortar bed and maintaining a 1/2-inch joint spacing. The cavity between the two rows of bricks can be filled with mortar as you go along, or can be poured full of grout after the job is complete.

Position tie-strips or reinforcing wire to tie the two rows of brick together. A few of the bricks must be cut to admit passage of the drain tubes, which should be placed near the bottom of the wall and exit just above grade level. The tubes extend back into gravel chimneys or pockets, and should be nominal 1/2-inch diameter plastic, copper or galvanized iron pipe. Metallic tubing such as is used in electrical work also will do the job.

To complete the wall, add a full cap of header bricks. If you can get at the back face, it is a good idea to brush on a coat of asphalt waterproofing to make the wall watertight. The finished retaining wall is shown in Fig. 9-19.

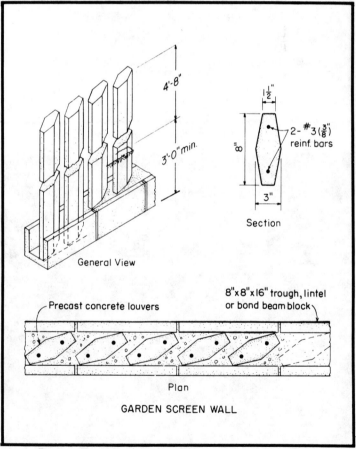

Fig. 9-17. General details of a louver-type garden screen wall.

327

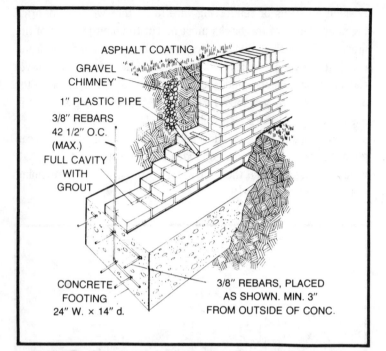

ASPHALT COATING

GRAVEL
CHIMNEY

1" PLASTIC PIPE

3/8" REBARS
42 1/2" O.C.
(MAX.)

FULL CAVITY
WITH
GROUT

CONCRETE
FOOTING
24" W. × 14" d.

3/8" REBARS, PLACED
AS SHOWN. MIN. 3"
FROM OUTSIDE OF CONC.

Fig. 9-18. Construction details of brick retaining wall.

Fig. 9-19. Completed brick retaining wall.

This particular wall is designed to be no more than 3 feet high. However, should you need a taller one, you can go to a different structural bonding pattern for far greater strength, sufficient to add on a foot or two to the same general wall design. Another possiblity is to go to a stronger structural bonding pattern and also increase the thickness of the wall to 12 or even 16 inches. The larger the wall, the more massive the footing should be, and the inclusion of additional reinforcing rods and ties would also be helpful. Note too that the same design can be used for a freestanding brick wall, though there are many others as well.

BRICK SCREEN WALL

The brick screen wall makes a handsome addition to a home, and can be built in low form to surround a patio or service yard, or in a tall form such as the one in Fig. 9-20. Though the structure shown is only a 4-inch wall, it has sufficient strength and stability to last for a good long time. However, the same project could be made in an 8-inch thickness, and reinforced with ties or reinforcing wire in

Fig. 9-20. Brick screen wall.

Fig. 9-21. General construction arrangement of brick screen wall. Courtesy of the Brick Institute of America.

either width. Ends and corners could be made in pilaster or column fashion for even greater strength, and also a somewhat different appearance.

The general layout for a brick screen wall is shown in Fig. 9-21. Start with a substantial footing placed at least 18 inches below grade level, or below frost line if that happens to be lower. When starting to lay up the masonry units, you have a choice. If you wish, you can start by laying a few courses of concrete block to bring the wall base up to grade level, thus saving a little bit of cost and labor. Or, you can start right off on the footing top with brick courses.

Either way, the first few courses should be laid in a solid running bond with a 1/2-overlap, to a height of perhaps 2 or 3 courses of brick above grade level. The actual dimensions of the brick laid in the screen portion of the wall, including those at the corners or ends, depend upon the overall size of the brick that you are using. The quoins will be somewhat less than full-length, and the small sections of brick are 1/2-bats laid endwise over the stretcher joints. If you break the bats from full brick, place the finished edge on the outside or most visible face of the wall, with the rough broken end on the inside. If appearance is important on both faces of the wall, saw the bricks in half with a power masonry saw, or have the job done by a masonry contractor.

The quoins and bats at the ends or corners should be sized so that the open spaces between bricks are all equal. Work out the proper dimensions and laying pattern with dry brick before you start actually laying up with mortar. Alternate the courses from openings/bats to full stretcher courses. Cap the wall with at least three and preferably more courses of stretchers to make a sturdy structure. A cap course of overhanging units could also be added if you wish.

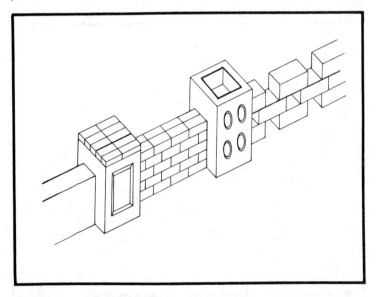

Fig. 9-22. Masonry fence consisting of concrete columns connected by masonry unit sections.

MASONRY FENCES

There are a great many possibilities for masonry fences, some of which consist entirely of masonry units of one sort or another, and others of which may include wood posts or rails.

For instance, one possibility is to pour a continuous footing below frost level along the line of the fence. Upon this footing, pour a series of heavy concrete columns spaced about 8 to 10 feet apart. These columns can be impressed with a design and capped with brick or stone. Between each column, lay up masonry unit wall sections of brick, block, stone or whatever else appeals, either in solid or screen fashion (Fig. 9-22).

Another method consists of pouring only a series of footings to support masonry columns at 8- or 10-foot intervals. Upon these footings build up brick, stone or block columns, leaving appropriate slots in which redwood, cedar or treated wood rails can be laid as the construction proceeds. The columns may be left hollow and later filled with dirt to serve as planters, or can be capped in whatever fashion you choose (Fig. 9-23).

Masonry fences can also be built without going to the trouble of pouring any footings at all (Fig. 9-24). Precast a series of concrete

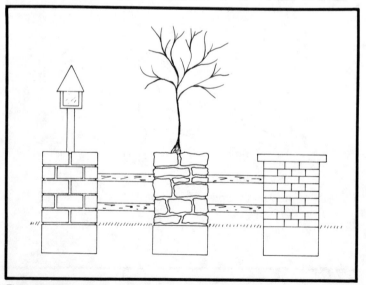

Fig. 9-23. Masonry fence consisting of masonry unit columns with wood fence rails between.

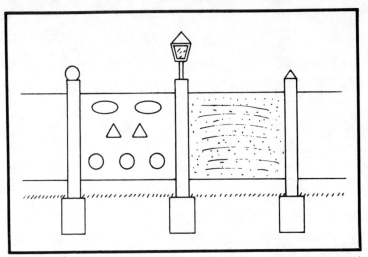

Fig. 9-24. Masonry fence consisting of cast concrete panels attached to wood posts.

slabs 1 1/2 inches thick, 2 to 3 feet wide and 4 to 6 feet long. Three #4 reinforcing bars should be set in each panel and protrude about 1 1/2 inches on each side. The form can be made from two-by-fours laid down flat, with holes drilled at appropriate points to center the three reinforcing bars in the thickness of the concrete and also to hold them in the proper positions. The concrete panels can be finished in any way you want, with textures, exposed aggregate, or even openings formed in place when the panel is poured.

After the panels cure, set them into fence posts by inserting the reinforcing bar pins into holes drilled into the posts. The posts are 4 inches by 4 inches or larger and of an appropriate height, set into holes in the ground. Make the holes at least 2 feet deep and pour concrete around the posts as you set them. Keep the posts and the panels, which are extremely heavy, well-supported and braced up as you erect the fence. Allow plenty of time for the concrete around the posts to cure before removing the braces.

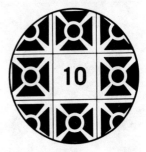

Edgings, Steps, Walks and Drives

Except in the heart of the city, every residence—whether modest or grand—sits on a parcel of land that can serve the household in both functional and decorative ways. Improving the grounds about a home with edgings and borders, walkways, steps, and driveways not only enhances the utility and appearance of the property, it also increases the monetary value. Because of masonry's attributes of versatility and permanence, it is ideally qualified for choice as the material to be used in the construction of these improvements. You yourself can use masonry to accomplish these improvements. Many of the projects that follow are extremely simple to build, while others are more difficult. But in all cases, there are several ways in which each project can be built or designed, and both methods and features can be changed about to suit the specific application and the fancy of the builder.

SPLIT BLOCK BORDER

The split block border shown in Fig. 10-1 is one of the simplest possible projects to put together. Cut a trench just a bit deeper than half the length of a block and of the same width. Spread a layer of loose soil or sand about 1 inch deep in the bottom of the trench. Set the first block in place at whatever angle you choose, and ram soil around it to hold it in place. Dampening the soil with water will allow

Fig. 10-1. Simple border made from split block.

you to pack it tighter. Working backward and away from the angle of the first block, set the subsequent blocks by leaning them into position and tapping the corners down into the loose soil at the bottom of the trench. Pack soil tightly around both faces of the blocks to wedge them firmly in place.

DRY BRICK EDGING

A dry-laid brick edging makes a handsome border and is just as easy to install as the one described above. One type of brick border is shown in Fig. 10-2. Lay out the line of the border strip and cut a trench, square-sided and flat-bottomed, the same width as the brick length and an inch deeper than the brick height. Spread a layer of fine damp sand into the bottom of the trench and level it to a smooth surface with a trowel. Make the layer a little more than an inch deep, so that when the bricks are set in they will rise a bit above grade level. After a short time they will settle down naturally.

Set the brick in place one by one and press them down firmly. As you lay them, keep the faces tight together and don't allow sand from the bed to work up between the bricks and force them apart. You can go around gentle curves with this border by just spreading the bricks along the curve into a fan pattern. Sharp corners will require cutting a few bricks into wedges. When the job is complete, press the soil or sod down firmly against the ends of the bricks to hold them tightly in place.

EDGING BLOCKS

There is a variety of ready-made masonry blocks that you can use for edging and border strips, and stone can also be effective. But there is another possibility, too. Make your own edging blocks from poured concrete, and finish them in some design of your own choos-

Fig. 10-2. Simple garden border of brick.

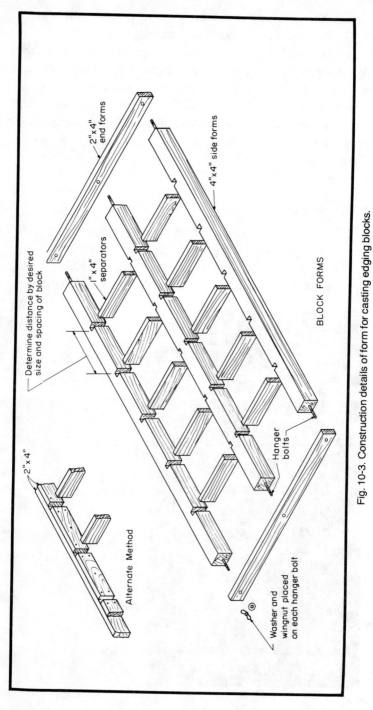

Fig. 10-3. Construction details of form for casting edging blocks.

2" x 4" end forms

4" x 4" side forms

Determine distance by desired size and spacing of block

1" x 4" separators

BLOCK FORMS

Hanger bolts

2" x 4"

Alternate Method

Washer and wingnut placed on each hanger bolt

338

ing. Figure 10-3 shows the details of a form for casting a dozen such blocks at one time. The form is easy to construct and is reusable for an indefinite number of pours. If you wish, you can make a smaller version which casts fewer blocks but is easier to build and to handle.

To make edging blocks, lay the form out on a watertight and oiled surface, or upon a sheet of plastic film laid on a smooth surface. Pour the assembled form full of concrete, strike and darby the surface and finish as usual. Allow the blocks plenty of time to cure before putting them to work.

Blocks which are cast upon a plastic sheet will have a glass-smooth finish. By laying plastic on a rough surface like gravel, you can impart a rough and nubby finish to the blocks. The various standard finishing procedures of floating and troweling can be used, or you can texture, stud the blocks with pebbles or mineral crystals, impress designs upon them, or whatever.

Though these blocks are intended for borders and edgings, there is no reason at all why they cannot be used for other projects as well. Such blocks are perfectly suitable for low walls, planters, and similar structures.

Cast Edging Strips

Another method of making edging is shown in Fig. 10-4. This form is designed for casting continuous 2-foot or 4-foot strips of concrete which can serve as boundaries for gardens, walkways, driveways or the like, or as wheel chocks for parking pads outdoors or in the garage. Instead of casting with standard portland cement, you can use white portland and add coloring pigments to make the strips more attractive. Or, you can paint them with concrete paint.

Short strips can be cast of plain concrete. Longer ones from 2 feet to 4 feet are stronger if a couple of lengths of reinforcing bar are included. You can also stick 2 or 3 long pins of reinforcing bar into the concrete so that they protrude from the bottom of the strip. The pins will bite into the ground and prevent the strips from sliding when they're bumped.

Working on the same principle, you can cast edging strips in about any form that you wish. Figure 10-5 shows a corrugated type of edging. This particular type is corrugated asbestos cement which is available in large sheets and in strips of a smaller, more convenient

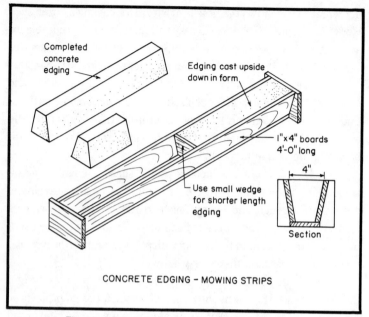

Fig. 10-4. Details for casting concrete edging strips.

size. However, you can cast your own by making a form of corrugated sheet metal and pouring it full of sand-mix concrete. Since the strips are rather delicate, short pieces work best. You can make edging strips in just about any shape just by assembling a suitable form.

DRY BRICK STEPPING STONES

Stepping stones make most attractive pathways and are extremely easy to build. A dry brick stepping stone project takes only a few hours to put together, is inexpensive, and provides a handsome effect. The diagram in Fig. 10-6 shows one general procedure for putting together brick stepping stones.

Choose a weather-resistant type of brick that appeals to you and lay a few of them out in a square or rectangular pattern of your choice. Measure the perimeter of the brick assembly, and build an appropriate number of wood frames whose inside dimensions match the outside dimensions of the brick assemblies. Redwood plank is most often used for this purpose, since it is impervious to moisture and insects and will remain intact indefinitely. However, other kinds of wood can be used if properly treated.

Cut square-sided holes in the sod or soil at the stepping stone locations, just large enough to accommodate the wood frames. The top of the frame should lie flush with grade level. The holes should be deep enough so that the brick tops will be even with the frame top edge when placed upon a bedding cushion about 1 to 1 1/2 inches

Fig. 10-5. Corrugated garden edging.

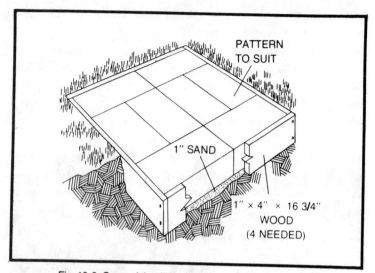

Fig. 10-6. General details for making brick stepping stone.

Fig. 10-7. Brick stepping stone pathway.

thick. Mix 3 parts clean sand with 1 part standard portland cement and spread this dry mixture in a 1- to 1 1/2-inch-deep cushion at the bottom of each hole. Lay the bricks into the frame and stomp them down firmly. If they are too low, remove them and add more sand-cement mix. The cushion will eventually pick up sufficient moisture to harden and lock the brick into place. Figure 10-7 shows a completed brick stepping stone path.

FLAGSTONE PATH

Flagstone paths are popular and can be used to good effect around the home. One of the easiest ways to make such a path is to cast the flagstone in place from concrete. Figure 10-8 shows the general layout.

Establish the line of the path first. Cut a series of flagstone patterns from cardboard or paper and lay them in place to form the pattern that you want. Alternatively, mark the edges of the path with string and cut the patterns directly into the soil or sod if you have a good eye for that sort of thing. Cut holes in the earth, square-sided and flat-bottomed, to match the shapes of the flagstones. A depth of 2 inches below grade level is satisfactory. Mix up a rich batch of concrete, dampen the bottoms of the holes, and pour the concrete

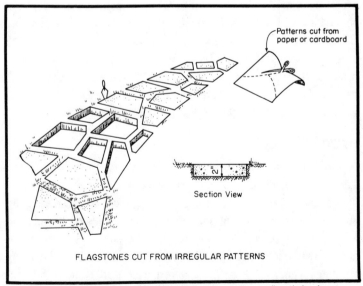

Patterns cut from paper or cardboard

Section View

FLAGSTONES CUT FROM IRREGULAR PATTERNS

Fig. 10-8. Arrangement for casting flagstone walkway directly in place.

directly in place. Deposit the concrete in the holes carefully and pack it tightly with no spill-over. Level the tops of the flagstones, let them set up for a bit, and then finish them with a wood float followed by one pass with a steel trowel. This will give a dense and weatherproof surface.

Another possibility is to precast a series of concrete flagstones. The easiest way to go about this is to design a section of path about 3 or 4 feet long and whatever width you wish, composed of a series of irregular flagstones properly fitted together. Make up a set of forms to match the flagstone shapes. Place the forms on a flat watertight surface, perhaps backed by a plastic sheet, and cast the flagstones and finish them. Make and install one section at a time until you have completed the entire length of the path. A thickness of 1 1/2 to 2 inches is sufficient for these flagstones, and a concrete mix of 1:2:2 1/4 is a good one for the purpose.

To lay the flagstones, you have two choices. One is to lay out the flagstones on the soil or sod and cut a series of holes to exactly accommodate them. Make the holes about 1 inch deeper than the thickness of the flagstones and lay a sand cushion to make up the difference. In a soft or moisture-laden area, make the holes even deeper so that you can put in a couple of inches of fine gravel as well. The second method is to dig out the entire pathway to a depth of about 4 inches for dry, solid soil or 6 to 8 inches for soft, moist soil. Cover the bottom of the excavation with a layer of gravel, well tamped. Follow this with a 2-inch layer of sand, wetted and tamped firm. Lay the flagstones in place, leaving a suitable gap between each to form the joint pattern. After the flagstones are laid, fill the joint gaps with sand and settle the sand with a mist of water.

FLAGSTONE WALK

When a walkway of more massive flagstones is desired, the system shown in Fig. 10-9 works out well. After the site is prepared by removing sod and adding a gravel or sand cushion, depending upon the moisture content and drainage capabilities of the soil, the form is set in place. Pour and screed the concrete in the usual fashion, allow it to set, and follow up with any one of the many finishing techniques. After the concrete has cured, remove the form and flip it end for end to gain additional pattern variety, and pour

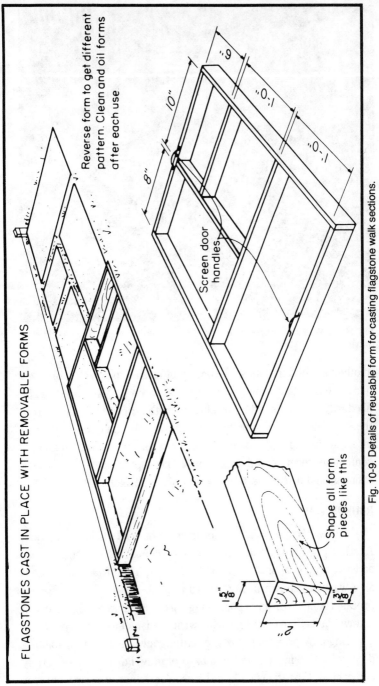

FLAGSTONES CAST IN PLACE WITH REMOVABLE FORMS

Reverse form to get different pattern. Clean and oil forms after each use

Screen door handles

8"

10"

6"

1'-0"

1'-0"

Shape all form pieces like this

5/8"

3/8"

2"

Fig. 10-9. Details of reusable form for casting flagstone walk sections.

Fig. 10-10. Walkway of precast exposed-aggregate concrete rounds bedded in gravel.

another set of flagstones. A coating of release oil on the form will make it easier to remove. The space between the flagstones can be filled with dirt and seeded, or plugged with strips of sod saved from the excavation, or filled with sand or gravel. Another possibility is to insert strips of redwood. In the latter case, you could also make a continuous form of redwood, mask the top to prevent concrete staining, and just leave the wood in place permanently.

ROUNDS WALK

Pathways and walks do not have to be made from flagstones. Another popular style is referred to as *rounds*, which can be made and arranged in an infinite number of ways. The broad and impressive walkway shown in Fig. 10-10 is made up of several sizes of precast rounds with an exposed-aggregate finish, randomly placed in a gravel bed. Constructing such a walkway is not a difficult project.

Begin by preparing the site, cutting away sod and topsoil to a depth of 4 to 6 inches. Put in a base of gravel or road-base, firmly compacted. Cast the rounds in the usual fashion and apply whatever finish appeals to you. Set the rounds in position and bed them firmly into the base. Moistening the base with water makes this job easier and aids in compaction and firm setting. Add border strips of redwood or some other treated wood if you wish, and fill in the remaining area with washed gravel.

DRY BRICK WALK

A brick walk makes a handsome addition to any home, and can be successfully laid without benefit of mortar. Best results are obtained, however, when a substantial border is installed on each side, as shown in Fig. 10-11. This border consists of a poured concrete footing topped with a row of brick mortared into place, but a solidly anchored redwood rail would work, too. In fact, any method that holds the edges of the block perfectly stable will suffice.

Make the excavation as required and put in a 4-inch layer of gravel topped with an inch of sand as shown. The layer of 15-pound roofing felt accomplishes three purposes. First, it is easier to work upon than the sand layer. Second, where ground moisture levels are high, the felt will prevent a growth of green algae from appearing on the brick. And last, the sand which is brushed into the joints between the bricks will stay put and not work down into the cushion.

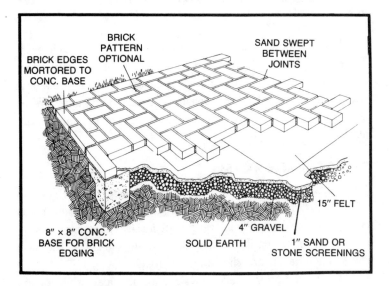

Fig. 10-11. General details for constructing a mortarless brick walk.

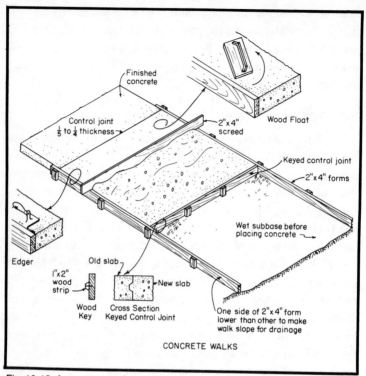

Fig. 10-12. Arrangements for making a typical poured concrete continuous slab walk.

Build the borders first in whatever manner you have chosen. Then fill in the middle with brick arranged in any pattern which appeals to you. Allow for a bit of a slope by slanting the entire walk in one direction or another, or crowning it slightly in the center. When all the bricks are set, sweep fine sand into the joints until they are filled.

POURED CONCRETE WALKS

Perhaps the most permanent and maintenance-free of all walks is the poured concrete variety. Basic construction is the same for all types, but the tremendous variety of finishes that are possible means that you can easily build a walk unlike anyone else's. The basic construction details are shown in Fig. 10-12.

A good width for a service walk is about 2 feet, while a good width for a main walk is 3 feet or more. A 6-gallon concrete mix will provide a durable and long-lived walk with a weather-resistant surface. Make the slab about 3 1/2 to 4 inches thick and reinforce with mesh. Groove the surface about 1 inch deep every 4 feet. Cut the trench to the same depth as the slab thickness if the soil is stable and well-drained. Otherwise, make the excavation deeper and put in a 4- to 6-inch layer of gravel or road-base, well compacted. In either case, dampen the surface thoroughly before pouring the concrete.

The process of making a concrete walk is the same as for any slab construction. Prepare the site, build the forms, mix and place the concrete, and screed the surface. Follow up with a darby. After the concrete has set up, do the finishing work, cut the grooves and edge the sides of the slab. Moist-curing should be done for at least 3 days, and preferably more. If the walk is poured in two or more separate operations, use keyed control joints between sections as shown in the diagram.

MORTARED WALKS

There are plenty of possibilities for making walkways of all sorts from various unit pieces tied together with either mortar or concrete. For instance, the mortared brick walkway works out very nicely. This requires making a concrete slab and then placing the bricks on the slab in a full mortar bed and with full mortar head and side joints. There are other construction methods as well; refer to the discussion on paving with brick and pavers in Chapter Eight.

Another possibility is to make a walk from mortared wood rounds or squares. Wood rounds are made by sawing short chunks from logs with a chainsaw. Squares are made in the same way, except that they are sawed from old railroad ties or any other treated rectangular beams. A length of 4 to 6 inches is sufficient for either type.

To make a mortared wood walkway, start by digging a trench about 2 inches deeper than the length of the rounds or squares. Place a 2-inch layer of sand or fine gravel at the bottom of the trench, and stand the rounds or squares on end on this cushion so that their tops are level with one another and even with the grade level, or a bit higher if you wish. If the units are stacked rather close together, pack a fairly wet and rich mortar all around them. If there are large

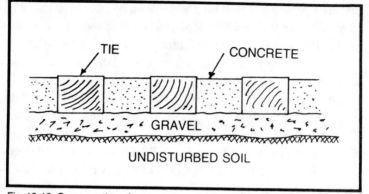

Fig. 10-13. Cross section of a concrete and railroad tie or wood beam walkway.

gaps between, pour concrete around the units. Smooth off the top of the concrete, and let the whole affair cure before walking upon it.

Another interesting walkway can be made with railroad ties and concrete. Prepare a trench for the walkway in the usual manner, with a sand or gravel cushion. Cut a suitable number of railroad ties into pieces the same length as the width of the walk. Lay out the pieces ladder-fashion, spaced one, two or three tie-widths apart, or whatever distance you like. Make a temporary form by nailing boards across the ends of the ties to enclose the spaces between the ties. Pour these spaces full with concrete and finish the surface however you wish. Or, fill the spaces partly full and lay brick or flat stone level with the tie tops. This works even better if you mortar the joints between the bricks or stones. Figure 10-13 will give you an idea of how this project goes together.

BRICK-AND-WOOD STEPS

Though most steps are of solid construction to avoid the danger of their later loosening up or settling out of shape, here is a combination which can be put together easily from brick and wood without benefit of concrete or mortar (Fig. 10-14).

The most critical part of the operation is making the excavation. The step form should be notched out of native soil with as little disruption as possible so as to maintain a solid base. Start at the bottom and work up the slope. Bed the heavy wood beams solidly in the earth, maintaining reasonable tread widths and riser heights. If you wish, you can drill holes through the beams and drive 4-foot lengths of 1/2-inch reinforcing bar through the wood and into the ground to peg the pieces securely in place. Lay the bricks dry in the usual fashion on a bed of sand or gravel; fit them tightly together and set them snugly between the beams. Pack soil or sod in tightly along the edges of the structure to help hold everything in place.

STEPS OF PRECAST ROUNDS

Another type of steps that can be used on long, gentle grades or to surmount fairly steep slopes consists of a series of precast concrete rounds. Cast a number of rounds of appropriate size and shape and of a minimum 2-inch thickness and with whatever finish you choose. Making the steps involves nothing more than placing the rounds so that they overlap each other slightly and step upward from unit to unit from bottom to top. They may be placed in a straight line, but are often more effective and have a more pleasing appearance

Fig. 10-14. Steps of wood beams and dry-laid brick.

Fig. 10-15. Steps, or inclined step-ramp, of precast exposed-aggregate concrete rounds.

when they traverse a slope as in Fig. 10-15. Make sure that the rounds are firmly bedded in the soil so that they don't wobble about, and are positioned in a way to minimize the effects of erosion and water drainage. For best results, carefully cut away the undisturbed native soil and fit each round closely to the face of the cut, packing damp earth all around.

POURED CONCRETE STEPS

Building poured concrete steps—probably the most rugged and durable of all types—is a matter of assembling a suitable form and then pouring and finishing the concrete. The basics of step construction were discussed earlier. Figure 10-16 gives the details for stepped ramp construction, most useful where the grades are gentle and the approach path fairly long. Such steps are easily formed and rest upon tamped earth. Any of the many concrete finishes can be applied, and the final results need not look like plain old concrete as Fig. 10-17 will attest. Here sizable stones were embedded in the concrete as the pouring proceeded. Note that there is no need for the steps and landings to follow any particular pattern; they may be spaced in whatever fashion seems appropriate.

BRICK STEPS

Brick steps have a great deal more charm and a more pleasing appearance in many residential applications than plain concrete steps. There are several methods for going about the building of brick steps. One calls for the construction of a heavy reinforced concrete footing covering the entire area beneath the steps. The steps themselves are built by laying up course after course of brick, stepping them up and back in appropriate tread/riser combinations until the final height is reached. This calls for a lot of brick, and a lot of labor. The construction process, however, is not particularly dif-

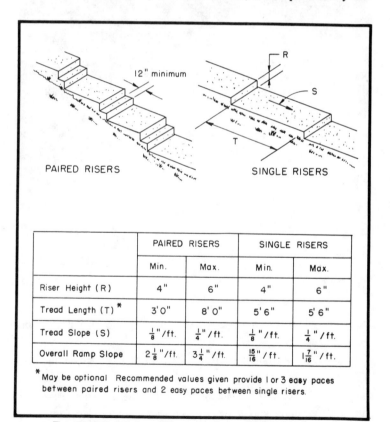

	PAIRED RISERS		SINGLE RISERS	
	Min.	Max.	Min.	Max.
Riser Height (R)	4"	6"	4"	6"
Tread Length (T) *	3' 0"	8' 0"	5' 6"	5' 6"
Tread Slope (S)	$\frac{1}{8}$"/ft.	$\frac{1}{4}$"/ft.	$\frac{1}{8}$"/ft.	$\frac{1}{4}$"/ft.
Overall Ramp Slope	$2\frac{1}{8}$"/ft.	$3\frac{1}{4}$"/ft.	$\frac{15}{16}$"/ft.	$1\frac{7}{16}$"/ft.

* May be optional Recommended values given provide 1 or 3 easy paces between paired risers and 2 easy paces between single risers.

Fig. 10-16. Details for making poured concrete stepped ramps.

Fig. 10-17. Modified stepped ramp of poured concrete and embedded cobble-stones.

ficult and only requires that good workmanship and careful assembly be practiced.

A second method is to first pour a set of steps from concrete. This can be done by erecting suitable forms and pouring against a tamped earth ramp, or by filling in much of the form's interior with rubble. The rubble should be covered with a smooth layer of gravel before pouring. The forms must be dimensioned, especially the tread width and riser height, with the fact in mind that bricks will be later installed upon the steps. The only finishing required on the concrete is a final wood floating, which should leave the surface level yet rough enough to afford a good key for the mortar between it and the brick.

After the poured concrete steps have been allowed to cure for at least a week, the process of bricklaying can begin. Start from the bottom and work up, pasting the bricks to the concrete with a 1/2-inch full bed of mortar. If your calculations were correct, no brick cutting will be necessary. And, incidentally, the same process can be employed to cover an existing set of concrete steps. The principal requirement is that the steps be thoroughly cleaned and washed down, even with muriatic acid if necessary. If the steps are quite smooth, they should be roughened up to provide better mortar keying. It is likely that some brick cutting will be necessary, but you

Fig. 10-18. Brick entry steps.

Fig. 10-19. Driveway of dry-laid brick.

may be able to make many adjustments by varying the width of the mortar joints.

Another construction method is to first pour a footing beneath the steps, and then build up a set of steps from concrete block. Again, the dimensions must be calculated nicely to avoid later problems when the brick veneering is applied. After the concrete block joints have cured satisfactorily, the brick veneer is applied in exactly the same fashion as on the poured concrete. Whichever method is used, the finished results will look something like the steps shown in Fig. 10-18.

DRY BRICK DRIVEWAY

Installing a dry-laid brick driveway is no minor project that can be completed in a weekend, but still is one of the easiest types to build. The mechanics of putting together the driveway shown in Fig. 10-19 are exactly the same as for the dry-laid brick walkway discussed earlier and diagrammed in Fig. 10-11. The only difference is the width. Drainage, however, is an even more important consideration in a driveway than in a walk. Make absolutely certain that there is sufficient slope in one direction or another to allow surface moisture to drain away rapidly.

In areas of severe winter weather, this type of driveway can become problematical because of the constant freeze-thaw action taking place beneath it. A gravel layer of 8 to 12 inches, instead of 4 as used in the walkway, would be of some help in this respect. However, a certain amount of instability is unavoidable unless the brick is laid upon a concrete base. On the other hand, repairs are far more easily made to this kind of driveway than any other. A small amount of routine maintenance each spring will keep its appearance up.

PAVER DRIVE

Paver driveways can also be laid dry, in the same fashion as the brick driveway discussed above. Better results, however, are obtained from laying the pavers on a solid foundation. There are two general types of paver driveways: solid, where the paver units tightly abut each other, and frequently interlock as well; and grid type, which is designed to allow grass to grow through the gridwork.

Construction details for building a driveway of pavers are similar to the standard paver laying practices treated in Chapter Eight. All that is necessary is to lay out the lines of the driveway, install a retaining border strip, and fill in the space with pavers.

CONCRETE BLOCK DRIVEWAY

You can make an informal driveway suitable for car traffic from standard concrete blocks (Fig. 10-20). The finished driveway has a pleasing appearance, is inexpensive to build and doesn't require a great amount of work.

Lay out the dimensions of the driveway to exactly accommodate a series of standard 8-inch by 8-inch by 16-inch stretcher blocks, along with the necessary half blocks, laid out in a running bond with 1/2-overlap. If you can, use the 3-core type of block. Excavate the driveway area to a depth of 12 inches and deposit a level and well-tamped 4-inch layer of road-base or fine gravel mixed

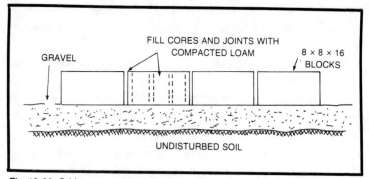

Fig. 10-20. Grid-type grass and masonry driveway made from standard concrete block laid with cores vertical, cross section.

with sand. Lay the block in place in a running bond, bottom down and with the cores exposed, leaving a 3/8-inch joint gap between all spaces. Set the blocks firmly in place and align them accurately.

When the blocks are laid, or at least a substantial portion of them, shovel good loam that is free from rocks and debris out onto the blocks and ram the cores full. Force loam into the joint cracks, filling them completely. Liberal spraying with a garden hose until the loam turns almost to mud will help pack and settle the soil tightly into the cores and joint cracks. Continue the process until no more loam can be rammed into place and the entire driveway is firm and well-set.

Loosen the top layer of soil with a garden rake and sow grass seed over the entire surface. Clover works particularly well and is fast-growing, but can also spread to other parts of the lawn in a short time. If this is a concern, choose another kind of seed. Keep traffic off the driveway until the grass has had a chance to establish itself. Be sure to give the new growth plenty of water; this will also further settle and firm up the blocks. Maintain the driveway as you would any lawn. If snowplowing is a necessity, however, plow with a blade equipped with shoes so that the edge of the blade does not catch on the blocks.

CONCRETE DRIVEWAY

A poured concrete driveway is unquestionably the toughest and most durable of all. It is highly weather-resistant, easily maintained, a snap to plow or shovel off and unaffected by gasoline or oil spills.

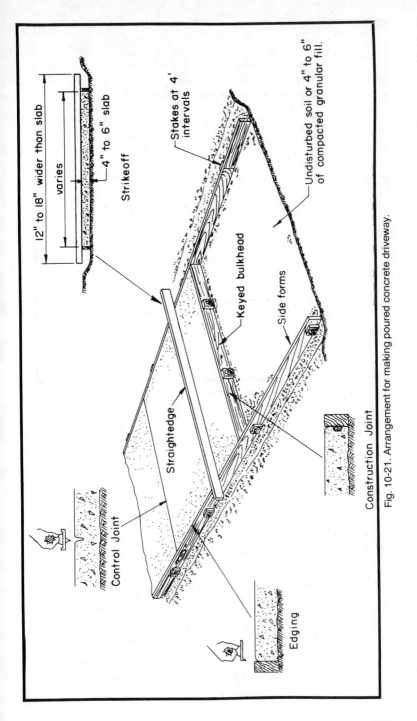

12" to 18" wider than slab

varies

4" to 6" slab

Strikeoff

Stakes at 4' intervals

Undisturbed soil or 4" to 6" of compacted granular fill.

Keyed bulkhead

Side forms

Straightedge

Control Joint

Construction Joint

Edging

Fig. 10-21. Arrangement for making poured concrete driveway.

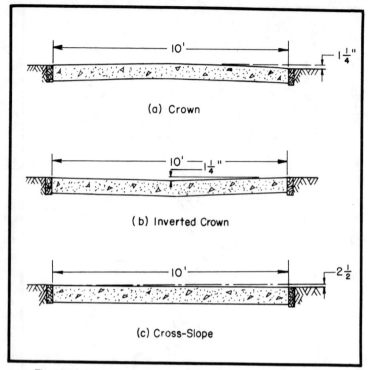

Fig. 10-22. Methods of providing drainage for concrete driveways.

Properly built, a concrete driveway will last indefinitely, and by virtue of coloring and/or the application of a decorative finish, it need not be just another drab slab. Nor is a poured concrete driveway difficult to build. Figure 10-21 shows a construction method for a typical driveway.

This type of driveway should be 10 to 14 feet wide; if curves are included, 14 feet should be the minimum width. A short driveway servicing a two-car garage can run from 16 to 24 feet wide. A long driveway to a two-car garage is normally single-width throughout the approach, then widens into a broad apron just in front of the garage. A thickness of 4 inches is considered adequate for residential driveways which receive only the lightweight traffic of passenger vehicles. However, if you anticipate heavyweight traffic in the nature of trucks and recreational vehicles, go to a 6-inch thickness. These thicknesses will actually be 3 1/2 and 5 1/2 inches, since stock two-by-fours and two-by-sixes will be used for the form rails. In either case, include a full layer of reinforcing mesh. The concrete can be poured directly onto firm native soil if the drainage is good. Otherwise, make the excavation deeper and include a well-compacted 4- to 8-inch layer of gravel or road-base beneath the concrete. Grooves or control joints should be installed about every 10 feet.

There are two regards in which many poured concrete driveways fall short of the mark. The first is in provision for proper drainage. Figure 10-22 shows some possibilities for good drainage.

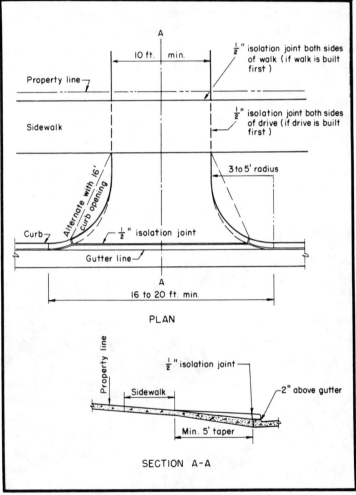

Fig. 10-23. Details of concrete driveway entrance.

The minimum slope for efficient drainage is 1/8 inch for every running foot. Side drainage is usually provided on a drive which is level lengthwise. If there is a lengthwise slope to the driveway, side drainage should be built in anyway for quick runoff unless the drive is quite steep. A lengthwise slope, incidentally, should always be away from the garage or parking area, and not towards it—for obvious reasons.

The second shortcoming is involved with the connection of the driveway's approach section to the street or service road. Sufficient space should be allowed to enable a driver to turn onto or out of the driveway with ease, and any dip should be gentle enough so there is no danger that the vehicle will bottom. Figure 10-23 shows the details of a typical driveway entrance. In general, the slope of the entrance should be no more than 14 percent, or 1 3/4 inches of rise for every running foot. The driveway itself should embody as gentle a slope as possible, with the parking area or a substantial length of driveway directly in front of the garage graded level.

The design and finish of the driveway should conform to the theme of the grounds and the architecture of the residence. If planning and execution are carried through properly, the driveway's overall effect will be complementary and not detractive. For example, Fig. 10-24 shows an exposed-aggregate driveway bordered

Fig. 10-24. Exposed-aggregate concrete driveway with brick borders.

with brick which harmonizes admirably with the building and grounds. Note the attractive low masonry walls which further enhance the area and join the masonry driveway visually to the masonry-veneered residence.

The actual construction of a poured concrete driveway follows the standard procedures for slab structures. Work out the dimensional details and the type of finish first. Prepare the site and construct the forms, adding a gravel layer if necessary. Dampen the surface and place the concrete, starting from the furthest point and working back; strike and darby. When the concrete has set up, perform the finish operations and moist-cure for 3 to 5 days.

Patios and Barbecues

One of the best additions that an owner can make to enhance the enjoyment and liveability of his home is a patio or barbecue or some combination of both. There is a strong tendency these days toward outdoor living of all sorts, and patios and barbecues have become almost standard fixtures in the residential good life. So much so, in fact, that complexes consisting of a barbecue or firepit, patio, and perhaps arbors, screen walls, seating arrangements and other accessory improvements have come to be known as outdoor rooms.

All of the structures mentioned above would be excellent projects for the do-it-yourselfer to undertake. Though many are far from simple, even a beginner can expect to turn in a creditable job by proceeding with care and patience. Large projects do take a considerable amount of time and labor, but the overall expense is comparatively low and the satisfaction derived from building and utilizing an important improvement to your property is substantial.

PRECAST SLAB PATIO

Of the many construction possibilities for a patio, the precast slab method is about the easiest and least expensive. Though a certain amount of yearly maintenance is necessary, especially in areas of severe winters or of heavy annual rainfall, the total of the returns that can be expected for a relatively small outlay of effort and money is great.

Fig. 11-1. Patio area laid out and excavated.

The precast slabs can be of any size and shape that you wish to make them. The slab surfaces can be plain or finished, colored or not, patterned or textured, and laid out in decorative patterns or mosaics, however you want. The smaller sizes, such as 1 foot or 2 feet square, are light in weight and easy to handle. The larger sizes, 1 foot by 2 feet or even 2 feet by 4 feet, are more difficult to lay but also more stable.

Small paving slabs can be cast 1 1/2 inches thick, but the larger ones should be on the order of 2 1/2 inches thick. Single or multisection forms can be easily made up from nominal one-by-two or two-by-two stock. Set the forms on a flat, watertight surface and fill them with concrete; later finish the pavers to your liking. If you prefer a perfectly smooth finish with practically no work, lay the forms on a flat surface covered with a sheet of construction plastic. Allow the paving slabs to cure for several days before laying them. You can make stronger units by embedding a couple of short pieces of rebar or a rectangle of remesh in the center of each slab.

The first step in building the patio is to stake out the area. Dimension the patio to suit the dimensions of the cast paving slabs. This is important when you purchase the slabs, so that you can avoid having to do any cutting. If you cast your own, you can cast whatever sizes and shapes are needed to fit the patio design. Stake out the corners of the patio and string a line between the stakes to serve as a guide. Excavate the interior to a depth equal to the paving slab thickness plus another 2 to 3 inches for a sand bed (Fig. 11-1). If the ground has a high moisture content for any length of time during the year, dig down another 6 inches for laying a bed of road-base.

Fill the excavation with 2 or 3 inches of damp sand. Rough-level the sand and compact it firmly with a tamper. Embed leveling boards of nominal two-by-two stock, redwood or some other treated wood, into the sand bed with the tops of the boards positioned the same distance below the desired finish grade of the patio as the thickness of the paving slabs (Fig. 11-2). The leveling boards should be set to a slant of 1/4 inch per running foot so that water will drain off the patio.

Pack sand in around the leveling boards and take a few quick sightings across the sand bed. If there are obvious low spots, add more sand. Then draw a long, straight-edged piece of lumber along the leveling boards just as though you were screeding concrete (Fig. 11-3). Compact the sand bed again with a tamper, and if necessary add another thin layer of sand until the entire bed is even with the leveling boards.

Begin at one corner or another and lay the paving slabs carefully in place (Fig. 11-4). Note that the leveling boards remain in place

Fig. 11-2. Sand cushion emplaced, leveling boards set and screed ready to go.

Fig. 11-3. Smoothing out the sand cushion.

permanently. Continue to lay the slabs; work by standing on the previously laid slabs rather than treading on the sand bed. The slabs can be butted tightly together, or may be arranged with a 1/4-inch or 3/8-inch separation between. Lay all of the slabs in place until the

Fig. 11-4. Placing the patio paving slabs in position on the sand cushion.

entire patio area is covered (Fig. 11-5). Set them firmly onto the sand bed, but don't stomp them down; let normal traffic and weathering firm them. When all the slabs are laid, brush fine sand into the joints. This may have to be repeated a couple of times, especially if a heavy rainstorm should occur before the sand has settled completely.

PAVER PATIO

Although there are more individual pieces to handle, a patio built from paver blocks is just as easy to make as a patio built from paving slabs, and the general procedure is the same for a sand-bed type of construction. However, the smaller paving blocks do not have the flotation of the large slabs and so they are more prone to get out of alignment and level, especially in localities with bad winter weather. A more permanent and maintenance-free patio can be built by laying the pavers on a concrete slab or asphalt cushion, just as might be done for a walkway or driveway. Further details can be found in Chapter Eight.

The first step in building a paver patio is to calculate the size and shape of the area in whole-number increments of the size of the pavers you plan to use. Dig out the topsoil and sod to a suitable depth to allow for the pavers and a subbase of crushed gravel (or roadbase) for drainage—if necessary—plus a concrete slab, asphalt cushion or sand cushion. Build a border strip first around the perimeter of the patio, setting retaining bricks or pavers in concrete to

Fig. 11-5. Slab patio nearing completion.

Fig. 11-6. Constructing redwood grid to serve as form for sectional poured concrete patio.

prevent the pavers from creeping. Lay the pavers in place with mortared, sand-filled, or dry cement/sand-filled joints, just as though you were building a driveway.

SECTIONED CONCRETE PATIO

The easiest way to make a poured concrete patio is to divide the area into small squares. This is particularly helpful in a do-it-yourself project because each square can be treated as a mini-project and the job can be accomplished piecemeal as time, funds and energy allow. In addition, the dividers between the concrete sections will act as control joints to make the patio durable and weatherable. This is an excellent project to make from prepackaged concrete mix, working with small batches in series, rather than having to mix and pour huge amounts of concrete at one time.

Dig out the sod and subsoil first and level the subbase to a reasonably smooth surface. Determine the proper finished grade level and build a continuous grid of construction-grade two-by-fours of redwood or some other treated wood to cover the entire area

370

(Fig. 11-6). This will serve as a form for the concrete, as well as a permanent decoration.

The forms shown here are supported above subgrade and held to finished grade level by steel pins secured to the lower portions of the form rails and driven into the ground for additional strength and support. This can be done with chunks of rebar fastened in place by U-clips. An alternative method is to support the frame with chunks of rock, brick or concrete blocks.

Make constant use of a spirit level as you build the form and check each section to make sure that it is perfectly square. Grade stakes and guidelines of string are helpful in monitoring the proper level. The entire patio should slope in the most convenient direction at a rate of 1/4 inch per running foot for good drainage. Figure 11-7 shows a typical patio grid form completed. Note the other masonry projects which are going on at the same time as the patio construction.

When the form is complete, fill the entire area with a bed of clean and rock-free sand to serve as a base for the concrete slab (Fig.

Fig. 11-7. Completed redwood grid form.

Fig. 11-8. Emplacing and leveling a layer of clean sand in the patio form.

11-8). Rough-level the sand in each section with a square-bladed shovel so that it is filled to within 2 inches of the form tops. Tamp the sand firm, soak with water and tamp again. Level the sand surface in each section with a short piece of board and bring the level up flat and smooth to a point 2 inches below finished grade level. If necessary, tamp again and relevel, adding sand as needed, until you have a smooth, even and densely compacted cushion.

The patio is now ready to receive concrete. You can order ready-mix and do the whole job at once; or rent a small power mixer, lay in the materials and mix your own. Or, you can do the entire project by mixing small batches of prepackaged concrete mix (Fig. 11-9). Even for large projects like this one, packaged mix is a perfectly reasonable approach for the home mechanic, and sometimes the best one.

As the batches are mixed, wheel them across elevated runways temporarily placed for the purpose (Fig. 11-10). Start your pouring at the farthest point. Dump concrete into each grid section (Fig.

11-11) until it is full. As soon as each section is filled, screed the surface and push the excess concrete into the next empty section.

Continue this process until all of the grid squares are filled, or until the time arrives when you have to commence finishing operations on previously poured sections. Unless you have a helper, this means that you will have to discontinue pouring after about 2 hours and start finishing. Float the concrete surface to begin with, and then go on to any other finish that you wish to apply. As soon as the

Fig. 11-9. Mixing prepackaged concrete in garden wheelbarrow.

Fig. 11-10. Wheeling mixed concrete across runways to the pouring point.

finishing operation is complete, carry out the curing procedures by covering with plastic film, spraying on curing compound or whatever else may be advisable. Then continue with the pouring operations elsewhere. Figure 11-12 shows a typical sectioned patio, the same one that appears unfinished in Fig. 11-7. Note the masonry screen wall at the rear, firepit covered with a removable redwood top in the patio center, flagstone retaining wall at the upper left, brick-veneered concrete block pool and fountain, and brick-veneered concrete block redwood-topped wall/benches which also serve as planters.

FREEFORM CONCRETE PATIO

The freeform poured-concrete patio is made in much the same manner as the sectional poured-concrete patio. The principal differences are that the patio is constructed in any desired random shape rather than as a rectangle, and that the thickness of the concrete is usually (though not always) greater. This patio is also poured in

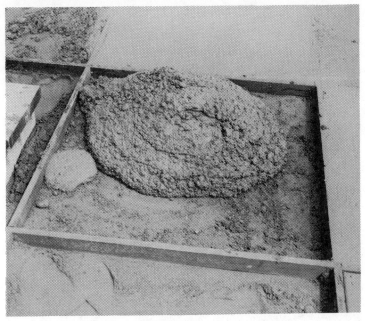

Fig. 11-11. Emplacing concrete mix in form sections.

Fig. 11-12. Completed sectional concrete patio with decorative gridwork.

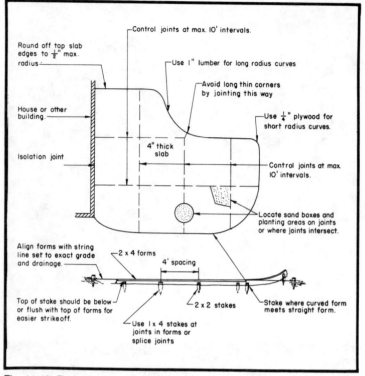

Fig. 11-13. Typical arrangement for building freeform poured concrete patio.

sections, with control joints placed on maximum 10-foot spacings. This insures that the patio surface will not buckle or crack. The control joints can be made from nominal 1-inch or 2-inch redwood or other treated wood, or from a special control joint material made for the purpose. Figure 11-13 shows a typical freeform poured-concrete patio layout.

Though small sections of poured concrete in a patio can be successfully made only 2 inches or so thick, large sections must be heavier. In either case, greater strength can be gained by the inclusion of squares of remesh within the sections. For most residential applications, a 4-inch thickness is perfectly adequate, but a 5-inch thickness gives extra insurance. A 6-inch thickness is more than necessary except in areas of extremely bad winter weather. Figure 11-14 shows how these slabs are formed up. The portions of the perimeter of a freeform patio that are straight can be formed up in the usual fashion. Those portions which are curved, however, must be handled a bit differently. Figure 11-15 shows some of the methods for forming curves. General construction methods are the same as for the sectional poured-concrete patio.

PATIO AND STEPS

Not all patios are plain and simple, nor need they be made only on one level. Patios can be built on two or several levels and may include one or more steps, as well as integrated seat/walls, retaining

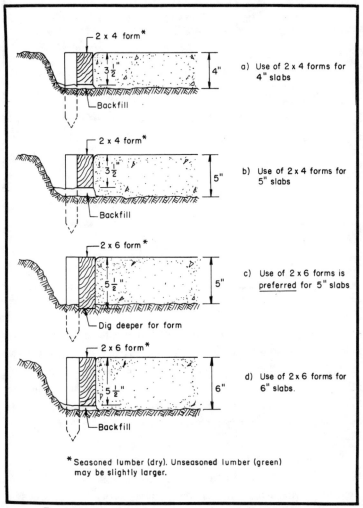

Fig. 11-14. Methods of forming for various slab thicknesses.

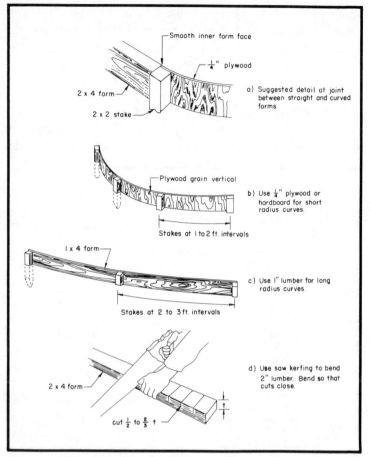

Smooth inner form face

¼" plywood

2 x 4 form

2 x 2 stake

a) Suggested detail at joint between straight and curved forms.

Plywood grain vertical

b) Use ¼" plywood or hardboard for short radius curves.

Stakes at 1 to 2 ft. intervals

1 x 4 form

c) Use 1" lumber for long radius curves.

Stakes at 2 to 3 ft. intervals

2 x 4 form

cut ½ to ⅔ t

d) Use saw kerfing to bend 2" lumber. Bend so that cuts close.

Fig. 11-15. Methods of constructing curved forms.

walls and other ancillary structures. In most cases the various elements are put together concurrently. An example of this is shown in Fig. 11-16, where a set of broad steps has been poured to reach the patio level. Note that the steps as well as the patio surface are embedded with fist-sized stones, lending a rough texture and a rustic appearance which overcomes the rather drab aspect of plain concrete.

Constructing a patio of this sort is largely a matter of building the correct forms upon a face of tamped subsoil, filled with a sand and/or gravel cushion. Install control joints every 10 feet or less in both directions, and isolation joints wherever the concrete slab

378

adjoins an existing structure or solid object. Lay in reinforcing mesh and emplace a 2-inch layer of concrete. Set the cobblestones in the concrete and pour the remaining mix to bring the slab up to grade level while at the same time exposing the tops of the cobblestones. Finish and cure in the usual manner.

BRICK PATIOS

Brick is another masonry unit which lends itself nicely to the building of patios. There are two general methods of doing the job. You can dry-lay the brick upon a sand cushion, or mortar them permanently to a poured concrete base. Either way, use brick pavers or other brick which is weatherable enough to withstand freeze-thaw action.

Figure 11-17 shows a typical mortarless patio. The perimeter of the patio is set upon a concrete footing, and the low walls serve both for seating and for retaining the bricks comprising the deck. Note the small barbecue built as an integral part of the seating wall.

The construction details for a mortarless brick patio are the same as for a mortarless brick driveway or walk. For construction details, refer to Fig. 10-11.

Fig. 11-16. Combined poured concrete patio and steps with embedded cobble-stones, rough-finished for informal appearance.

Fig. 11-17. Dry-laid brick patio. Seating wall and barbecue grille are mortared.

Though a mortared brick patio is more difficult to build, it is more durable and has a lower maintenance factor. Figure 11-18 shows a typical mortared brick patio. In this instance, a fairly high protective wall is a part of the project, while the barbecue unit is built upon its own footing and joined to the patio by its smaller front section. A single step is included as a part of the design, but the patio could just as well be built perfectly flat or in multiple levels.

The general construction details of a mortared brick patio are shown in Fig. 11-19. A stepped patio is made in the same fashion, except that the construction commences at the step section. Pour a footing for the step first, and then build the steps with a suitable number of courses of brick. The steps will also serve as a retainer for the paving brick. Fill the interior space with rubble, tamped and compacted firmly and topped with a sand or gravel layer, and pour a concrete slab base for the paving brick. The location of the step section and the finished height of the concrete slab must be accurately calculated so that the paving surface will be at the correct level and so that the pavers will fit properly within the border. When the concrete slab has cured, lay the paving brick in mortar in the usual manner. As with all patios, the surface should be sloped at a rate of 1/4 inch per running foot to provide proper drainage.

FIREPITS

Outdoor barbecuing is a favorite leisure pastime, and the building of a permanent cooking structure has become quite popular. Of such structures, by far the easiest to build is a barbecue pit or firepit.

One simple method is to precast a concrete ring that looks like a big donut. Make an inner form of 1/4-inch plywood or Masonite in a circle about 2 or 3 feet in diameter. Brace the form with an X of two-by-fours inside the circle. Make the outer form the same way,

Fig. 11-18. Mortared brick patio.

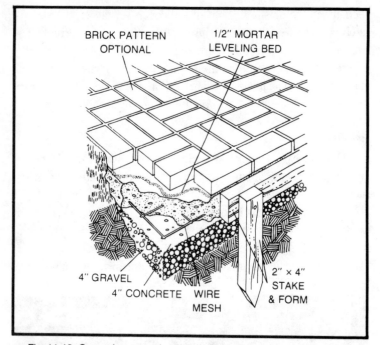

Fig. 11-19. General construction details for making mortared brick patio.

but 1 or 1 1/2 feet larger in diameter. Tie the two forms together with spreaders, set the form on a flat surface and pour full of concrete. Finish and cure in the usual manner. The finished thickness should be about 3 1/2 inches.

Make a circular excavation at the firepit site just big enough to hold the ring and about 8 inches deep. Cover the area which will lie directly beneath the ring with a bed of crushed stone and set the ring in place. Fill the hole in the center with pebbles.

To carry this project along further, you could then arrange a seating wall of poured concrete or brick in a half circle to one side of the firepit, with a brick or flagstone patio on the other side. Fill the space between the seating wall and the firepit with a layer of white marble pebbles or chips.

Another way to build a firepit is to dig a circular hole deep enough to contain a 2-inch concrete slab plus two courses of firebrick stood on end. About 3 feet is a good diameter for the circle. Pour a 2-inch concrete footing in the bottom of the hole, directly on the

382

earth and reinforced with a square of rebar. Mortar together two courses of firebrick stood on end to form the walls of the firepit (refer to Fig. 11-7, center). At grade level, lay another ring of brick—flat—and mortared to the firebrick. Every other brick should have provision for a 1-inch diameter hole running completely through the firepit wall. This allows draft for the fire. Lay another course of brick around the top and finish (refer to Fig. 11-12, center). Fill the firepit to just below finished grade level with fine pea gravel. This material, which can be easily removed and replaced when necessary, makes it possible to do pit baking.

DRY-LAID BRICK BARBECUE

If you would like to have a full-fledged outdoor barbecue fireplace but don't want to fuss around with mortaring, don't despair. By following the general idea shown in Fig. 11-20 you can build up a barbecue unit in an hour's time and never touch a mason's trowel.

Fig. 11-20. Mortarless brick barbecue, here set upon dry-laid brick patio.

Fig. 11-21. Mortared brick barbecue.

The most important point to remember in building a mortarless barbecue is that the base must be absolutely dead level. Naturally a concrete slab, well reinforced, makes the best base because of its strength and stability. Failing that, however, you can dig away sod and topsoil until you strike solid and undisturbed subsoil, the more heavily compacted the better. Lay in a 2-inch bed of sand; then wet and compact it. Now you can start stacking brick. Better yet, build your barbecue unit on an existing patio, which can be either dry-laid (as in the case of the unit shown), mortared brick or paver, or a concrete slab.

There is no particular trick to building up this barbecue unit. Crisscross the bricks in stacks and wherever you can, run tie-bricks to help hold the stacks together. The bricks may sometimes get out of alignment as you stack them. To realign them, hold and push down slightly on the upper bricks while tapping the offending brick with a rubber mallet. Don't belt them with a hammer, they'll break. The brick, of course, should be either firebrick or SW grade for complete weatherability. Purchase the grille racks first so that you can size the opening to them.

MORTARED BRICK BARBECUE

Obviously a mortared brick barbecue is a good deal more substantial than one made up of stacked bricks. There are a thousand different styles and designs, but Fig. 11-21 shows a fairly simple construction which is easily put together. The bond pattern is an easy one, with the cutting of bricks held to a minimum.

The details of construction are shown in Fig. 11-22. This is an exceptionally sturdy unit which is set upon a 1 1/2-foot-deep footing as shown. After the footing has cured, the first five courses of brick are laid to form a perimeter of the unit. After allowing sufficient time

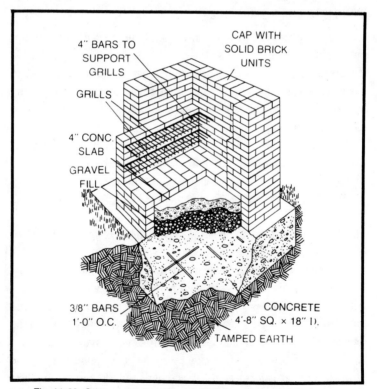

Fig. 11-22. General construction details for mortared brick barbecue.

for the mortar joints to cure, partially fill the opening with gravel and top with another concrete slab to a finished level even with the top of the fourth course of brick. After this slab has cured, pave it with brick.

Continue the construction of the barbecue by following the bond pattern shown in the diagram, working your way up course by course. As with most brick laying, the corners and ends should be built up three or four courses high first, and then filled, in a continuous succession until you reach the cap course. As the construction proceeds, insert pins at appropriate points in the fresh mortar joints. These pins will later serve to support the grilles. The pins can be made from 4-inch lengths of rebar or 4-inch bolts. A 1/4-inch diameter is adequate, though you can go to a 3/8-inch diameter if you wish. When the structure is complete, tool all the joints, brush away excess mortar crumbs and clean up as necessary.

CONCRETE BLOCK FIREPLACE

A barbecue unit need not necessarily be built from brick, nor need it be quite as sturdy as the one previously discussed. A plain concrete block barbecue is just as functional and quite a bit easier to construct. The appearance can be improved if you wish by using pattern blocks, painting or covering the basic structure with a veneer of tile, brick or a coating of stucco.

Start the proceedings by building a slab foundation as shown in Fig. 11-23. If frost is no problem, dig the sod and topsoil away to a depth of about 2 1/2 inches and set the two-by-four form in place. Make the finished level of the slab surface 1 inch above ground level. If the drainage is bad in that particular area or if the winters are harsh, dig down deeper and put in a 6-inch layer of crushed gravel or road-base and place the form on top of this.

Figure 11-24 shows the construction details of the fireplace unit itself. Lay the concrete block with full mortar bedding and line the interior of the firebox with firebrick laid on edge and fully mortared into place. For added strength, place rebar in the end cores and the outside chimney cores and bed them in sand, gravel or concrete. Concrete, of course, makes for the strongest construction. The grate holders are mortared into place between the second and third courses of block, set in appropriate positions. The metal grate

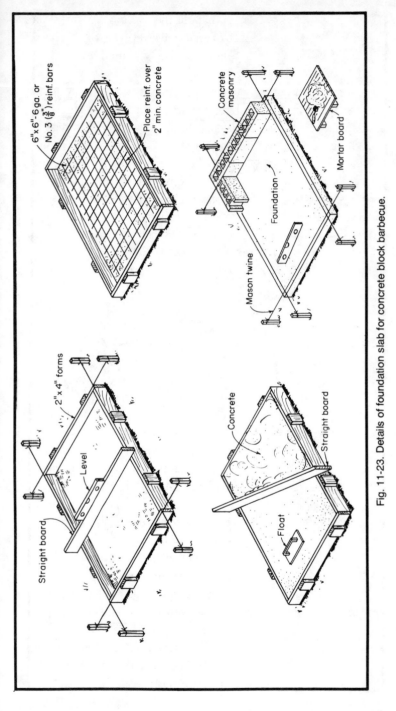

Fig. 11-23. Details of foundation slab for concrete block barbecue.

6" x 6" - 6 ga. or No. 3 ($\frac{3}{8}$") reinf. bars

Place reinf. over 2" min. concrete

Concrete masonry

Foundation

Mason twine

Mortar board

2" x 4" forms

Level

Straight board

Concrete

Straight board

Float

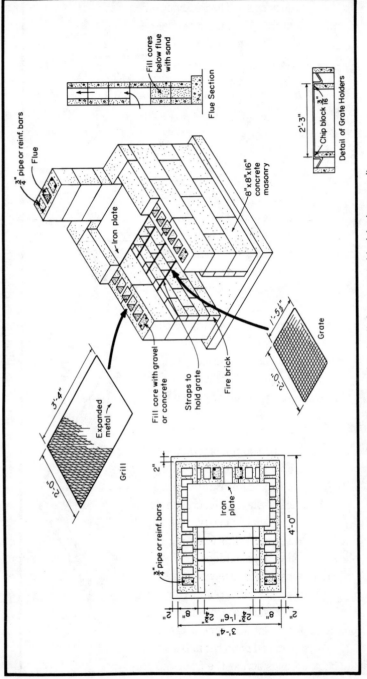

Fig. 11-24. General construction details for concrete block barbecue unit.

simply rests upon them. The iron baffle plate, which also doubles as a warmer, is mortared into place between the top course of block and the cap course. The removable grille of expanded metal lies across the top course of block. An alternative method which would produce a more finished appearance is to provide a second series of grate holders set across the top course of blocks and mortared into place under a continuation of the cap course. The grill would then be sized to lie upon those holders.

PATIO-BARBECUE-SHELTER COMBINATION

A patio and barbecue unit can be much more than just that, especially when combined with an underground shelter or room. For those who are interested in tackling a complex building project, Fig. 11-25 will be of interest. This is a general layout only, and there are a great many different possibilities as far as design specifics are concerned. Poured concrete, concrete block, brick, stone, pavers or various other materials could be used in a limitless number of arrangements. The room beneath the patio might serve as a fallout shelter, a storm shelter or tornado cellar, a storage area for garden tools and equipment or even a workshop. By combining methods, techniques and materials discussed in this book along with a liberal dash of imagination and enthusiasm, you can work up home outdoor masonry projects to suit any need or whim.

OUTDOOR ROOMS

The outdoor room is another example of a complex masonry project which takes some imagination and a degree of skill to properly design and execute. But there is little question that it is within the capabilities of the do-it-yourselfer, or that it does a great deal to enhance a residential property. One such outdoor room is shown in Fig. 11-12. Figure 11-18 featuring the mortared brick patio also shows a portion of an outdoor room of rather simple overall design, including steps, protective wall and barbecue unit. Another outdoor room of larger size and greater complexity is shown in Fig. 11-26. Construction of such a project is a major undertaking, but if approached a little bit at a time it can be done readily enough.

This is a raised patio set into a slope. The starting point is with the large footings across the front and at the far end to support the

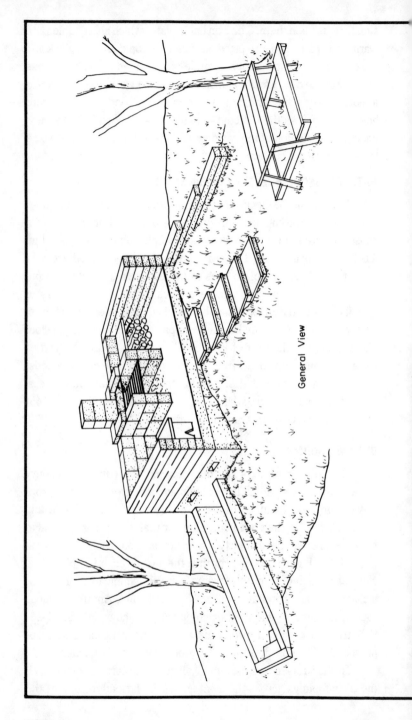

General View

390

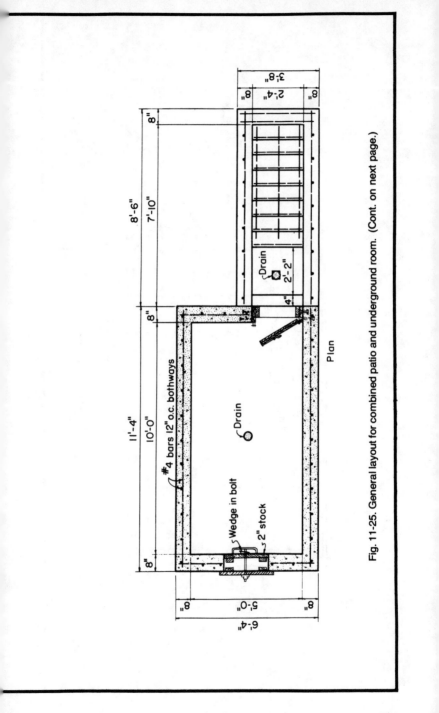

Fig. 11-25. General layout for combined patio and underground room. (Cont. on next page.)

Plan

#4 bars 12" o.c. bothways

Wedge in bolt

2" stock

Drain

Drain

8'-6"
7'-10"

11'-4"
10'-0"

8"

8"

8"

8"

6'-4"
5'-0"

8"

3'-8"
2'-4"
8"

2'-2"

4"

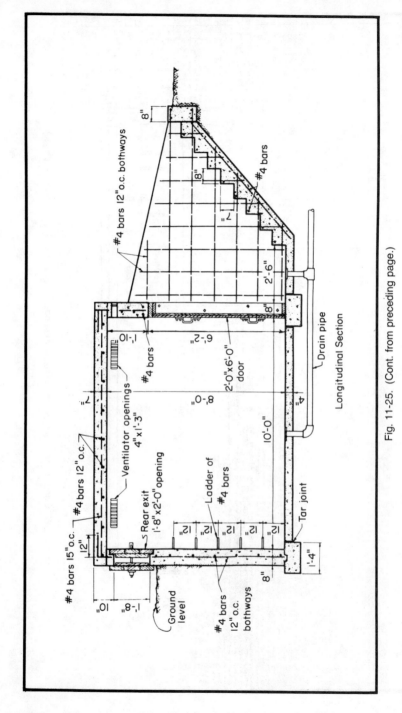

Fig. 11-25. (Cont. from preceding page.)

Longitudinal Section

#4 bars 12" o.c. bothways

#4 bars

#4 bars 12" o.c.

#4 bars

Ventilator openings 4"x1'-3"

Rear exit 1'-8"x2'-0" opening

#4 bars 15" o.c.

Ladder of #4 bars

2'-0"x6'-0" door

Drain pipe

Tar joint

Ground level

#4 bars 12" o.c. bothways

Fig. 11-26. Complete outdoor room.

393

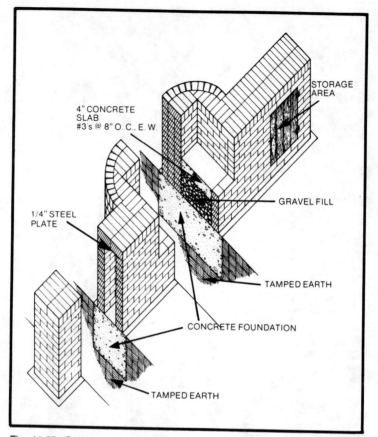

Fig. 11-27. Cutaway of portions of the outdoor room showing construction method.

front retaining wall, steps leading down to the lawn, and the combined barbecue unit and storage wall at the far end. The front wall and barbecue wall serve as borders for the brick paving, and another small border is set in concrete along the back edges of the patio. The interior portion is filled with tamped earth and rubble and topped with a poured concrete slab. The brick pavers are then laid upon this slab. The lamp post column and the barbecue and storage end wall can be built after the brick paving is laid, or concurrently with construction of the front and end walls. Figure 11-27 shows a cutaway drawing of the basic construction details of the project. By following the same general principles and making design changes as necessary, you can evolve a similar project that will fulfill your specific needs.

APPENDIX

Table A-1. ACI Recommended Maximum Permissible Water-Cement Ratios for Different Types of Structures and Degrees of Exposure.

Type of structures	Exposure conditions**					
	Severe wide range in temperature or frequent alternations of freezing and thawing (air-entrained concrete only)			Mild temperature rarely below freezing, or rainy, or arid (gallons/sack)		
	In air	At water line or within range of fluctuating water level or spray		In air	At water line or within range of fluctuating water level or spray	
		In fresh water	In sea water or in contact with sulfates†		In fresh water	In sea water or in contact with sulfates†
A. Thin sections such as reinforced piles and pipe	5.5	5	4.5	6	5.5	4.5
B. Bridge decks	5	5	4.5	5.5	5.5	5
C. Thin sections such as railings, curbs, sills, ledges, ornamental or architectural concrete, and all sections with less than 1-in. concrete cover over reinforcement	5.5	...	...	6	5.5	...
D. Moderate sections, such as retaining walls, abutments, piers, girders, beams	6	5.5	5	††	6	5
E. Exterior portions of heavy (mass) sections	6.5	5.5	5	††	6	5
F. Concrete deposited by tremie under water	...	5	5	...	5	5
G. Concrete slabs laid on the ground	6	...	...	††	...	...
H. Pavements	5.5	...	...	6	...	...
I. Concrete protected from the weather, interiors of buildings, concrete below ground	††	...	...	††	...	...
J. Concrete which will later be protected by enclosure or backfill but which may be exposed to freezing and thawing for several years before such protection is offered	6	...	...	††	...	...

*Adapted from Recommended Practice for Selecting Proportions for Concrete (ACI 613-54).
**Air-entrained concrete should be used under all conditions involving severe exposure and may be used under mild exposure conditions to improve workability of the mixture.
†Soil or groundwater containing sulfate concentrations of more than 0.2 per cent. For moderate sulfate resistance, the tricalcium aluminate content of the cement should be limited to 5 per cent, and for high-sulfate resistance to 5 per cent. At equal cement contents, air-entrained concrete is significantly more resistant to sulfate attack than non-air-entrained concrete.
††Water-cement ratio should be selected on basis of strength and workability requirements, but minimum cement content should not be less than 470 lb. per cubic yard.

Table A-2. Age-Compressive Strength Relationship for Types I and III Air-entrained Portland Cement.

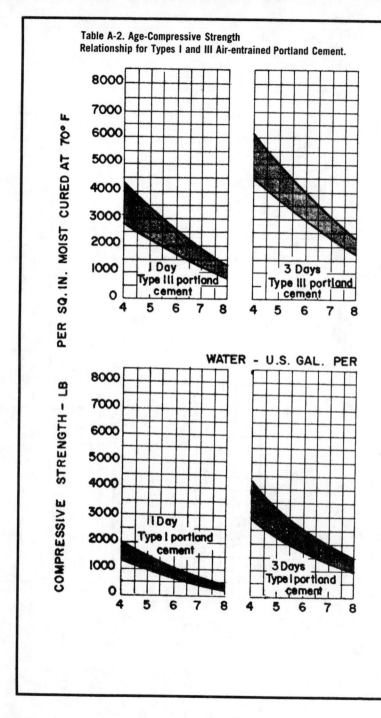

COMPRESSIVE STRENGTH – LB PER SQ. IN. MOIST CURED AT 70° F

WATER – U.S. GAL. PER

1 Day Type III portland cement

3 Days Type III portland cement

1 Day Type I portland cement

3 Days Type I portland cement

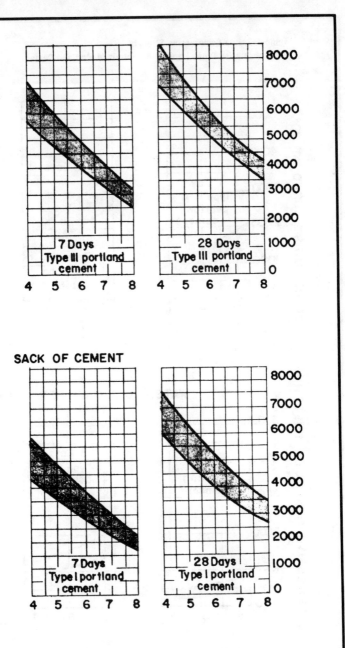

SACK OF CEMENT

(Cont on next page.)

399

Table A-2. Age-Compressive Strength Relationship for Types I and III Air-entrained Portland Cement.

(Cont. from previous page.)

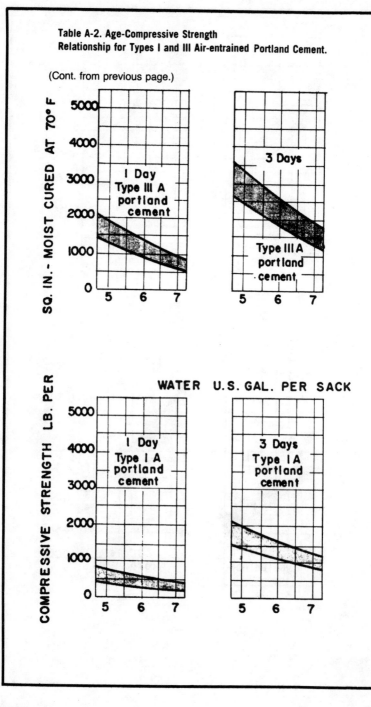

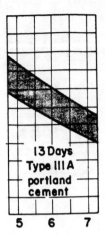

13 Days
Type III A
portland
cement

5 6 7

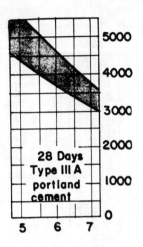

28 Days
Type III A
portland
cement

5000
4000
3000
2000
1000
0

5 6 7

OF CEMENT

13 Days
Type I A
portland
cement

5 6 7

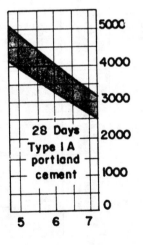

28 Days
Type I A
portland
cement

5000
4000
3000
2000
1000
0

5 6 7

Table A-3. Suggested Trial Mixes for Non-Air-Entrained Concrete of Medium Consistency with 3- to 4-Inch Slump.

Water-cement ratio Gal per sack	Maximum size of aggregate inches	Air content (entrapped air) per cent	Water gal per cu yd of concrete	Cement sacks per cu yd of concrete	With fine sand—fineness modulus = 2.50		
					Fine aggregate per cent of total aggregate	Fine aggregate lb per cu yd of concrete	Coarse aggregate lb per cu yd of concrete
4.5	3/8	3	46	10.3	50	1240	1260
	1/2	2.5	44	9.8	42	1100	1520
	3/4	2	41	9.1	35	960	1800
	1	1.5	39	8.7	32	910	1940
	1½	1	36	8.0	29	880	2110
5.0	3/8	3	46	9.2	51	1330	1260
	1/2	2.5	44	8.8	44	1180	1520
	3/4	2	41	8.2	37	1040	1800
	1	1.5	39	7.8	34	990	1940
	1½	1	36	7.2	31	960	2110
5.5	3/8	3	46	8.4	52	1390	1260
	1/2	2.5	44	8.0	45	1240	1520
	3/4	2	41	7.5	38	1090	1800
	1	1.5	39	7.1	35	1040	1940
	1½	1	36	6.5	32	1000	2110

6.0	3/8	3	46	7.7	53	1440	1260
	1/2	2.5	44	7.3	46	1290	1520
	3/4	2	41	6.8	39	1130	1800
	1	1.5	39	6.5	36	1080	1940
	1 1/2	1	36	6.0	33	1040	2110
6.5	3/8	3	46	7.1	54	1480	1260
	1/2	2.5	44	6.8	46	1320	1520
	3/4	2	41	6.3	39	1190	1800
	1	1.5	39	6.0	37	1120	1940
	1 1/2	1	36	5.5	34	1070	2110
7.0	3/8	3	46	6.6	55	1520	1260
	1/2	2.5	44	6.3	47	1360	1520
	3/4	2	41	5.9	40	1200	1800
	1	1.5	39	5.6	37	1150	1940
	1 1/2	1	36	5.1	34	1100	2110
7.5	3/8	3	46	6.1	55	1560	1260
	1/2	2.5	44	5.9	48	1400	1520
	3/4	2	41	5.5	41	1240	1800
	1	1.5	39	5.2	38	1190	1940
	1 1/2	1	36	4.8	35	1130	2110
8.0	3/8	3	46	5.7	56	1600	1260
	1/2	2.5	44	5.5	48	1440	1520
	3/4	2	41	5.1	42	1280	1800
	1	1.5	39	4.9	39	1220	1940
	1 1/2	1	36	4.5	35	1160	2110

*See footnote at end of table.

(Cont. on next page.)

(Cont. from previous page).

Table A-3. Suggested Trial Mixes for Non-Air-Entrained Concrete of Medium Consistency with 3- to 4-Inch Slump—continued.

Water-cement ratio Gal per sack	With average sand—fineness modulus = 2.75			With coarse sand—fineness modulus = 2.90		
	Fine aggregate percent of total aggregate	Fine aggregate lb per cu yd of concrete	Coarse aggregate lb per cu yd of concrete	Fine aggregate percent of total aggregate	Fine aggregate lb per cu yd of concrete	Coarse aggregate lb per cu yd of concrete
4.5	52	1310	1190	54	1350	1150
	45	1170	1450	47	1220	1400
	37	1030	1730	39	1080	1680
	34	980	1870	36	1020	1830
	32	960	2030	33	1000	1990
5.0	54	1400	1190	56	1440	1150
	46	1250	1450	48	1300	1400
	39	1110	1730	41	1160	1680
	36	1060	1870	38	1100	1830
	34	1040	2030	35	1080	1990
5.5	55	1460	1190	57	1500	1150
	47	1310	1450	49	1360	1400
	40	1160	1730	42	1210	1680
	37	1110	1870	39	1150	1830
	35	1080	2030	36	1120	1990
6.0	56	1510	1190	57	1550	1150
	48	1360	1450	50	1410	1400
	41	1200	1730	43	1250	1600
	38	1150	1870	39	1190	1830
	36	1120	2030	37	1160	1990

6.5	57	1550	1190	58	1590	1150
	49	1390	1450	51	1440	1400
	42	1240	1730	43	1290	1680
	39	1190	1870	40	1230	1830
	36	1150	2030	37	1190	1990
7.0	57	1590	1190	59	1630	1150
	50	1430	1450	51	1480	1400
	42	1270	1730	44	1320	1680
	39	1220	1870	41	1260	1830
	37	1180	2030	38	1220	1990
7.5	58	1630	1190	59	1670	1150
	50	1470	1450	52	1520	1400
	43	1310	1730	45	1370	1600
	40	1260	1870	42	1300	1830
	37	1210	2030	39	1250	1990
8.0	58	1670	1190	60	1710	1150
	51	1520	1450	53	1560	1400
	44	1350	1730	45	1400	1680
	41	1290	1870	42	1330	1830
	38	1250	2030	39	1280	1990

*Increase or decrease water per cubic yard by 3 per cent for each increase or decrease of 1 in. in slump, then calculate quantities by absolute volume method. For manufactured fine aggregate, increase percentage of fine aggregate by 3 and water by 17 lb. per cubic yard of concrete. For less workable concrete, as in pavements, decrease percentage of fine aggregate by 3 and water by 8 lb. per cubic yard of concrete.

Table A-4. Suggested Trial Mixes for Air-Entrained Concrete of Medium Consistency with 3- to 4-Inch Slump.

Water-cement ratio Gal per sack	Maximum size of aggregate inches	Air Content (entrapped air) per cent	Water gal per cu yd of concrete	Cement sacks per cu yd of concrete	With fine sand—fineness modulus = 2.50		
					Fine aggregate per cent of total aggregate	Fine aggregate lb per cu yd of concrete	Coarse aggregate lb per cu yd of concrete
4.5	3/8	7.5	41	9.1	50	1250	1260
	1/2	7.5	39	8.7	41	1060	1520
	3/4	6	36	8.0	35	970	1800
	1	6	34	7.8	32	900	1940
	1 1/2	5	32	7.1	29	870	2110
5.0	3/8	7.5	41	8.2	51	1330	1260
	1/2	7.5	39	7.8	43	1140	1520
	3/4	6	36	7.2	37	1040	1800
	1	6	34	6.8	33	970	1940
	1 1/2	5	32	6.4	31	930	2110
5.5	3/8	7.5	41	7.5	52	1390	1260
	1/2	7.5	39	7.1	44	1190	1520
	3/4	6	36	6.5	38	1090	1800
	1	6	34	6.2	34	1010	1940
	1 1/2	5	32	5.8	32	970	2110
6.0	3/8	7.5	41	6.8	53	1430	1260
	1/2	7.5	39	6.5	45	1230	1520
	3/4	6	36	6.0	38	1120	1800
	1	6	34	5.7	35	1040	1940
	1 1/2	5	32	5.3	32	1010	2110

6.5	3/8	7.5	41	6.3	54	1460	1260
	1/2	7.5	39	6.0	45	1260	1520
	3/4	6	36	5.5	39	1150	1800
	1	6	34	5.2	36	1080	1940
	1 1/2	5	32	4.9	33	1040	2110
7.0	3/8	7.5	41	5.9	54	1500	1260
	1/2	7.5	39	5.6	46	1300	1520
	3/4	6	36	5.1	40	1180	1800
	1	6	34	4.9	36	1100	1940
	1 1/2	5	32	4.6	33	1060	2110
7.5	3/8	7.5	41	5.5	55	1530	1260
	1/2	7.5	39	5.2	47	1330	1520
	3/4	6	36	4.8	40	1210	1800
	1	6	34	4.5	37	1140	1940
	1 1/2	5	32	4.3	34	1090	2110
8.0	3/8	7.5	41	5.1	55	1560	1260
	1/2	7.5	39	4.9	47	1360	1520
	3/4	6	36	4.5	41	1240	1800
	1	6	34	4.3	37	1160	1940
	1 1/2	5	32	4.0	34	1110	2110

*See footnote at end of table.

(Cont. on next page).

Table A-4. Suggested Trial Mixes for Air-Entrained
Concrete of Medium Consistency with 3- to 4-Inch Slump.

(Cont. from previous page).

Water-cement ratio Gal per sack	With average sand—fineness modulus = 2.75			With coarse sand—fineness modulus = 2.90		
	Fine aggregate percent of total aggregate	Fine aggregate lb per cu yd of concrete	Coarse aggregate lb per cu yd of concrete	Fine aggregate percent of total aggregate	Fine aggregate lb per cu yd of concrete	Coarse aggregate lb per cu yd of concrete
4.5	53	1320	1190	54	1360	1150
	44	1130	1450	46	1180	1400
	38	'040	1730	39	1090	1680
	34	970	1870	36	1010	1830
	32	950	2030	33	990	1990
5.0	54	1400	1190	56	1440	1150
	46	1210	1450	47	1260	14000
	39	1110	1730	41	1160	1630
	36	1040	1870	37	1080	1830
	33	1010	2030	35	1050	1990
5.5	55	1460	1190	57	1500	1150
	46	1260	1450	48	1310	1400
	40	1160	1730	42	1210	1680
	37	1080	1870	38	1120	1830
	34	1050	2030	35	1090	1990
6.0	56	1500	1190	57	1540	1150
	47	1300	1450	49	1350	1400
	41	1190	1730	42	1240	1680
	37	1110	1870	39	1150	1830
	35	1090	2030	36	1130	1990

6.5	56	1530	1190	58	1570	1150
	48	1330	1450	50	1380	1400
	41	1220	1730	43	1270	1680
	38	1150	1870	39	1190	1830
	36	1120	2030	37	1160	1990
7.0	57	1570	1190	58	1610	1150
	49	1370	1450	50	1420	1400
	42	1250	1730	44	1300	1680
	38	1170	1870	40	1210	1830
	36	1140	2030	37	1180	1990
7.5	57	1600	1190	59	1640	1150
	49	1400	1450	51	1450	1400
	43	1280	1730	44	1330	1680
	39	1210	1870	41	1250	1830
	37	1170	2030	38	1210	1990
8.0	58	1630	1190	59	1670	1150
	50	1430	1450	51	1480	1400
	43	1310	1730	44	1360	1680
	40	1230	1870	41	1270	1830
	37	1190	2030	38	1230	1990

*Increase or decrease water per cubic yard by 3 per cent for each increase or decrease of 1 in. in slump, then calculate quantities by absolute volume method For manufactured fine aggregate, increase percentage of fine aggregate by 3 and water by 17 lb. per cubic yard of concrete. For less workable concrete, as in pavements decrease percentage of fine aggregate by 3 and water by 8 lb. per cubic yard of concrete.

Table A-5. Approximate Mixing Water Requirements for Different Slumps and Maximum Sizes of Aggregates.

Maximum size of aggregate, in.	Air-entrained concrete				Approximate amount of entrapped air, per cent	Non-air-entrained concrete		
	Recommended average total air content, per cent†	Slump, in.				Slump, in.		
		1 to 2	3 to 4	5 to 6		1 to 2	3 to 4	5 to 6
		Water, gal. per cu.yd. of concrete**				Water, gal. per cu.yd. of concrete**		
⅜	7.5	37	41	43	3.0	42	46	49
½	7.5	36	39	41	2.5	40	44	46
¾	6.0	33	36	38	2.0	37	41	43
1	6.0	31	34	36	1.5	36	39	41
1½	5.0	29	32	34	1.0	33	36	38
2	5.0	27	30	32	0.5	31	34	36
3	4.0	25	28	30	0.3	29	32	34
6	3.0	22	24	26	0.2	25	28	30

*Adapted from Recommended Practice for Selecting Proportions for Concrete (ACI 613–54).
**These quantities of mixing water are for use in computing cement factors for trial batches. They are maximums for reasonably well-shaped angular coarse aggregates graded within limits of accepted specifications.
†Plus or minus 1 per cent.

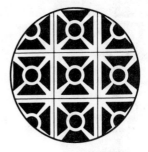

INDEX

412

415